ASE Test Preparation

Medium/Heavy Duty Truck Technician Certification Series

Electronic Diesel Engine Diagnosis Specialist (L2)

DELMAR
CENGAGE Learning·

Australia • Brazil • Japan • Korea • Mexico • Singapore • Spain • United Kingdom • United States

DELMAR
CENGAGE Learning

ASE Test Preparation: Medium/Heavy Duty Truck Technician Certification Series, Electronic Diesel Engine Diagnosis Specialist (L2)

Vice President, Technology and Trades Professional Business Unit: Gregory L. Clayton

Director, Professional Transportation Industry Training Solutions: Kristen L. Davis

Product Manager: Katie McGuire

Director of Marketing: Beth A. Lutz

Senior Marketing Manager: Jennifer Barbic

Senior Production Director: Wendy Troeger

Production Manager: Sherondra Thedford

Content Project Management: PreMediaGlobal

Senior Art Director: Benj Gleeksman

Section Opener Image: © Baloncici/www.shutterstock.com

For product information and technology assistance, contact us at
Cengage Learning Customer & Sales Support, 1-800-354-9706

For permission to use material from this text or product, submit all requests online at **www.cengage.com/permissions.**
Further permissions questions can be emailed to **permissionrequest@cengage.com.**

ISBN-13: 978-1-133-28046-0

ISBN-10: 1-133-28046-3

Delmar Cengage Learning
5 Maxwell Drive
Clifton Park, NY 12065-2919
USA

Cengage Learning is a leading provider of customized learning solutions with office locations around the globe, including Singapore, the United Kingdom, Australia, Mexico, Brazil, and Japan. Locate your local office at: **international.cengage.com/region**.

Cengage Learning products are represented in Canada by Nelson Education, Ltd.

For more information on transportation titles available from Delmar, Cengage Learning, please visit our website at **www.trainingbay.cengage.com**.

For more learning solutions, please visit our corporate website at **www.cengage.com**.

Notice to the Reader

Publisher does not warrant or guarantee any of the products described herein or perform any independent analysis in connection with any of the product information contained herein. Publisher does not assume, and expressly disclaims, any obligation to obtain and include information other than that provided to it by the manufacturer. The reader is expressly warned to consider and adopt all safety precautions that might be indicated by the activities described herein and to avoid all potential hazards. By following the instructions contained herein, the reader willingly assumes all risks in connection with such instructions. The publisher makes no representations or warranties of any kind, including but not limited to, the warranties of fitness for particular purpose or merchantability, nor are any such representations implied with respect to the material set forth herein, and the publisher takes no responsibility with respect to such material. The publisher shall not be liable for any special, consequential, or exemplary damages resulting, in whole or part, from the readers' use of, or reliance upon, this material.

Printed in the United States of America
1 2 3 4 5 6 7 16 15 14 13 12

Table of Contents

Delmar, a part of Cengage Learning, is very pleased that you have chosen to use our ASE Test Preparation Guide to help prepare yourself for the Electronic Diesel Engine Diagnosis Specialist (L2) ASE certification examination. This guide is designed to help prepare you for your actual exam by providing you with an overview and introduction of the testing process, introducing you to the task list for the Electronic Diesel Engine Diagnosis Specialist (L2) certification exam, giving you an understanding of what knowledge and skills you are expected to have in order to successfully perform the duties associated with each task area, and providing you with several preparation exams designed to emulate the live exam content in hopes of assessing your overall exam readiness.

If you have a basic working knowledge of the discipline you are testing for, you will find this book is an excellent guide, helping you understand the "must know" items needed to successfully pass the ASE certification exam. This manual is not a textbook. Its objective is to prepare the individual who has the existing requisite experience and knowledge to attempt the challenge of the ASE certification process. This guide cannot replace the hands-on experience and theoretical knowledge required by ASE to master the vehicle repair technology associated with this exam. If you are unable to understand more than a few of the preparation questions and their corresponding explanations in this book, it could be that you require either more shop-floor experience or further study.

This book begins by providing an overview of, and introduction to, the testing process. This section outlines what we recommend you do to prepare, what to expect on the actual test day, and overall methodologies for your success. This section is followed by a detailed overview of the ASE task list to include explanations of the knowledge and skills you must possess to successfully answer questions related to each particular task. After the task list, we provide six sample preparation exams for you to use as a means of evaluating areas of understanding, as well as areas requiring improvement in order to successfully pass the ASE exam. Delmar is the first and only test preparation organization to provide so many unique preparation exams. We enhanced our guides to include this support as a means of providing you with the best preparation product available. Section 6 of this guide includes the answer keys for each preparation exam, along with the answer explanations for each question. Each answer explanation also contains a reference back to the related task or tasks that it assesses. This will provide you with a quick and easy method for referring back to the task list whenever needed. The last section of this book contains blank answer sheet forms you can use as you attempt each preparation exam, along with a glossary of terms.

OUR COMMITMENT TO EXCELLENCE

Thank you for choosing Delmar, Cengage Learning for your ASE test preparation needs. All of the writers, editors, and Delmar staff have worked very hard to make this test preparation guide second to none. We feel confident that you will find this guide easy to use and extremely beneficial as you prepare for your actual ASE exam.

Delmar, Cengage Learning has sought out the best subject matter experts in the country to help with the development of *ASE Test Preparation: Medium/Heavy Duty Truck Technician Certification Series, Electronic Diesel Engine Diagnosis Specialist (L2), 1ˢᵗ Edition*. Preparation questions are authored and then reviewed by a group of certified subject-matter experts to ensure the highest level of quality and validity to our product.

If you have any questions concerning this guide or any guide in this series, please visit us on the web at **http://www.trainingbay.cengage.com**.

For web-based online test preparation for ASE certifications, please visit us on the web at **http://www.techniciantestprep.com/** to learn more.

ABOUT THE SERIES ADVISOR

Brian (BJ) Crowley has experienced several different aspects of the diesel industry over the past ten years. Now a diesel technician in the oil and gas industry, BJ owned and operated a diesel repair shop where he repaired heavy, medium, and light trucks—in addition to agricultural and construction equipment. He earned an associate's degree in diesel technology from Elizabethtown Community and Technical College and is an ASE Master certified medium/heavy truck technician.

The History and Purpose of ASE

ASE began as the National Institute for Automotive Service Excellence (NIASE). It was founded as a non-profit, independent entity in 1972 by a group of industry leaders with the single goal of providing a means for consumers to distinguish between incompetent and competent technicians. It accomplishes this goal through the testing and certification of repair and service professionals. Though it is still known as the National Institute for Automotive Service Excellence, it is now called "ASE" for short.

Today, ASE offers more than 40 certification exams in automotive, medium/heavy duty truck, collision repair and refinish, school bus, transit bus, parts specialist, automobile service consultant, and other industry-related areas. At this time, there are more than 385,000 professionals nationwide with current ASE certifications. These professionals are employed by new car and truck dealerships, independent repair facilities, fleets, service stations, franchised service facilities, and more.

ASE's certification exams are industry-driven and cover practically every on-highway vehicle service segment. The exams are designed to stress the knowledge of job-related skills. Certification consists of passing at least one exam and documenting two years of relevant work experience. To maintain certification, those with ASE credentials must be re-tested every five years.

While ASE certifications are a targeted means of acknowledging the skills and abilities of an individual technician, ASE also has a program designed to provide recognition for highly qualified repair, support, and parts businesses. The Blue Seal of Excellence Recognition Program allows businesses to showcase their technicians and their commitment to excellence. One of the requirements of becoming Blue Seal recognized is that the facility must have a minimum of 75 percent of their technicians ASE certified. Additional criteria apply, and program details can be found on the ASE website.

ASE recognized that educational programs serving the service and repair industry also needed a way to be recognized as having the faculty, facilities, and equipment to provide a quality education to students wanting to become service professionals. Through the combined efforts of ASE, industry, and education leaders, the non-profit organization entitled the National Automotive Technicians Education Foundation (NATEF) was created in 1983 to evaluate and recognize academic programs. Today more than 2,000 educational programs are NATEF certified.

For additional information about ASE, NATEF, or any of their programs, the following contact information can be used:

National Institute for Automotive Service Excellence (ASE)

101 Blue Seal Drive S.E.

Suite 101

Leesburg, VA 20175

Telephone: 703-669-6600

Fax: 703-669-6123

Website: **www.ase.com**

Overview and Introduction

Participating in the National Institute for Automotive Service Excellence (ASE) voluntary certification program provides you with the opportunity to demonstrate you are a qualified and skilled professional technician who has the "know-how" required to successfully work on today's modern vehicles.

EXAM ADMINISTRATION

> *Note:* After November 2011, ASE will no longer offer paper and pencil certification exams. There will be no Winter testing window in 2012, and ASE will offer and support CBT testing exclusively starting in April 2012.

ASE provides computer-based testing (CBT) exams, which are administered at test centers across the nation. It is recommended that you go to the ASE website at *http://www.ase.com* and review the conditions and requirements for this type of exam. There is also an exam demonstration page that allows you to personally experience how this type of exam operates before you register.

CBT exams are available four times annually, for two-month windows, with a month of no testing in between each testing window:

- January/February – Winter testing window
- April/May – Spring testing window
- July/August – Summer testing window
- October/November – Fall testing window

Please note, testing windows and timing may change. It is recommended you go to the ASE website at *http://www.ase.com* and review the latest testing schedules.

UNDERSTANDING TEST QUESTION BASICS

ASE exam questions are written by service industry experts. Each question on an exam is created during an ASE-hosted "item-writing" workshop. During these workshops, expert service representatives from manufacturers (domestic and import), aftermarket parts and equipment manufacturers, working technicians, and technical educators gather to share ideas and convert them into actual exam questions. Each exam question written by these experts must then survive review by all members of the group. The questions are designed to address the practical application of repair and diagnosis knowledge and skills practiced by technicians in their day-to-day work.

After the item-writing workshop, all questions are pre-tested and quality-checked on a national sample of technicians. Those questions that meet ASE standards of quality and accuracy are included in the scored sections of the exams; the "rejects" are sent back to the drawing board or discarded altogether.

Depending on the topic of the certification exam, you will be asked between 40 and 80 multiple-choice questions. You can determine the approximate number of questions you can expect to be asked during the Electronic Diesel Engine Diagnosis Specialist (L2) certification exam by reviewing the task list in Section 4 of this book. The five-year recertification exam will cover this same content; however, the number of questions for each content area of the recertification exam will be reduced by approximately one-half.

> *Note:* Exams may contain questions that are included for statistical research purposes only. Your answers to these questions will not affect your score, but since you do not know which ones they are, you should answer all questions in the exam.

Using multiple criteria, including cross-sections by age, race, and other background information, ASE is able to guarantee that exam questions do not include bias for or against any particular group. A question that shows bias toward any particular group is discarded.

TEST-TAKING STRATEGIES

Before beginning your exam, quickly look over the exam to determine the total number of questions that you will need to answer. Having this knowledge will help you manage your time throughout the exam to ensure you have enough available to answer all of the questions presented. Read through each question completely before marking your answer. Answer the questions in the order they appear on the exam. Leave the questions blank that you are not sure of and move on to the next question. You can return to those unanswered questions after you have finished the others. These questions may actually be easier to answer at a later time once your mind has had additional time to consider them on a subconscious level. In addition, you might find information in other questions that will help you recall the answers to some of them.

Multiple-choice exams are sometimes challenging because there are often several choices that may seem possible, or partially correct, and therefore it may be difficult to decide on the most appropriate answer choice. The best strategy, in this case, is to first determine the correct answer before looking at the answer options. If you see the answer you decided on, you should still be careful to examine the other answer options to make sure that none seems more correct than yours. If you do not know or are not sure of the answer, read each option very carefully and try to eliminate those options that you know are incorrect. That way, you can often arrive at the correct choice through a process of elimination.

If you have gone through the entire exam, and you still do not know the answer to some of the questions, *then guess*. Yes, guess. You then have at least a 25 percent chance of being correct. While your score is based on the number of questions answered correctly, any question left blank, or unanswered, is automatically scored as incorrect.

There is a lot of "folk" wisdom on the subject of test taking that you may hear about as you prepare for your ASE exam. For example, there are those who would advise you to avoid response options that use certain words such as *all, none, always, never, must,* and *only,* to name a few. This, they claim, is because nothing in life is exclusive. They would advise you to choose response options that use words that allow for some exception, such as *sometimes, frequently, rarely, often, usually, seldom,* and *normally*. They would also advise you to avoid the first and last option (A or D) because exam writers, they feel, are more comfortable if they put the correct answer in the middle (B or C) of the choices. Another recommendation often offered is to select the option that is either shorter or longer than the other three choices because it is more likely to be correct. Some would advise you to never change an answer since your first intuition is usually correct. Another area of "folk" wisdom focuses specifically on any repetitive patterns created by your question responses (e.g., A, B, C, A, B, C, A, B, C).

Many individuals may say that there are actual grains of truth in this "folk" wisdom, and whereas with some exams, this may prove true, it is not relevant in regard to the ASE certification exams. ASE validates all exam questions and test forms through a national sample of technicians, and only those questions and test forms that meet ASE standards of quality and accuracy are included in the scored sections of the exams. Any biased questions or patterns are discarded altogether, and therefore, it is highly unlikely you will experience any of this "folk" wisdom on an actual ASE exam.

PREPARING FOR THE EXAM

Delmar, Cengage Learning wants to make sure we are providing you with the most thorough preparation guide possible. To demonstrate this, we have included hundreds of preparation questions in this guide. These questions are designed to provide as many opportunities as possible to prepare you to successfully pass your ASE exam. The preparation approach we recommend and outline in this book is designed to help you build confidence in demonstrating what task area content you already know well while also outlining what areas you should review in more detail prior to the actual exam.

We recommend that your first step in the preparation process should be to thoroughly review Section 3 of this book. This section contains a description and explanation of the type of questions you'll find on an ASE exam.

Once you understand how the questions will be presented, we then recommend that you thoroughly review Section 4 of this book. This section contains information that will help you establish an understanding of what the exam will be evaluating, and specifically, how many questions to expect in each specific task area.

As your third preparatory step, we recommend you complete your first preparation exam, located in Section 5 of this book. Answer one question at a time. After you answer each question, review the answer and question explanation information located in Section 6. This section will provide you with instant response feedback, allowing you to gauge your progress, one question at a time, throughout this first preparation exam. If after reading the question explanation you do not feel you understand the reasoning for the correct answer, go back and review the task list overview (Section 4) for the task that is related to that question. Included with each question explanation is a clear identifier of the task area that is being assessed (e.g., Task A.1). If at that point you still do not feel you have a solid understanding of the material, identify a good source of information on the topic, such as an educational course, textbook, or other related source of topical learning, and do some additional studying.

After you have completed your first preparation exam and have reviewed your answers, you are ready to complete your next preparation exam. A total of six practice exams are available in Section 5 of this book. For your second preparation exam, we recommend that you answer the questions as if you were taking the actual exam. Do not use any reference material or allow any interruptions in order to get a feel for how you will do on the actual exam. Once you have answered all of the questions, grade your results using the Answer Key in Section 6. For every question that you gave an incorrect answer to, study the explanations to the answers and/or the overview of the related task areas. Try to determine the root cause for missing the question. The easiest thing to correct is learning the correct technical content. The hardest things to correct are behaviors that lead you to an incorrect conclusion. If you knew the information but still got the question incorrect, there is likely a test-taking behavior that will need to be corrected. An example of this would be reading too quickly and skipping over words that affect your reasoning. If you can identify what you did that caused you to answer the question incorrectly, you can eliminate that cause and improve your score.

Here are some basic guidelines to follow while preparing for the exam:

- Focus your studies on those areas you are weak in.
- Be honest with yourself when determining if you understand something.
- Study often but for short periods of time.
- Remove yourself from all distractions when studying.
- Keep in mind that the goal of studying is not just to pass the exam; the real goal is to learn.
- Prepare physically by getting a good night's rest before the exam, and eat meals that provide energy but do not cause discomfort.
- Arrive early to the exam site to avoid long waits as test candidates check in.
- Use all of the time available for your exams. If you finish early, spend the remaining time reviewing your answers.
- Do not leave any questions unanswered. If absolutely necessary, guess. All unanswered questions are automatically scored as incorrect.

Here are some items you will need to bring with you to the exam site:

- A valid government or school-issued photo ID
- Your test center admissions ticket
- A watch (not all test sites have clocks)

> *Note:* Books, calculators, and other reference materials are not allowed in the exam room. The exceptions to this list are English-Foreign dictionaries or glossaries. All items will be inspected before and after testing.

WHAT TO EXPECT DURING THE EXAM

When taking a CBT exam, as soon as you are seated in the testing center, you will be given a brief tutorial to acquaint you with the computer-delivered test prior to taking your certification exam(s). The CBT exams allow you to select only one answer per question. You can also change your answers as many times as you like. When you select a second answer choice, the CBT will automatically unselect your first answer choice. If you want to skip a question to return to later, you can utilize the "flag" feature, which will allow you to quickly identify and review questions whenever you are ready. Prior to completing your exam, you will also be provided with an opportunity to review your answers and address any unanswered questions.

TESTING TIME

Each individual ASE CBT exam has a fixed time limit. Individual exam times will vary based upon exam area, and will range anywhere from a half hour to two hours. You will also be given an additional 30 minutes beyond what is allotted to complete your exams to ensure you have adequate time to perform all necessary check-in procedures, complete a brief CBT tutorial, and potentially complete a post-test survey.

You can register for and take multiple CBT exams during one testing appointment. The maximum time allotment for a CBT appointment is four and a half hours. If you happen to register for so many exams that you will require more time than this, your exams will be scheduled into multiple appointments. This could mean that you have testing on both the morning and afternoon of the same day, or they could be scheduled on different days, depending on your personal preference and the test center's schedule.

It is important to understand that if you arrive late for your CBT test appointment, you will not be able to make up any missed time. You will only have the scheduled amount of time remaining in your appointment to complete your exam(s).

Also, while most people finish their CBT exams within the time allowed, others might feel rushed or not be able to finish the test, due to the implied stress of a specific, individual time limit allotment. Before you register for the CBT exams, you should review the number of exam questions that will be asked along with the amount of time allotted for that exam to determine whether you feel comfortable with the designated time limitation or not.

As an overall time management recommendation, you should monitor your progress and set a time limit you will follow with regard to how much time you will spend on each individual exam question. This should be based on the total number of questions you will be answering.

Also, it is very important to note that if for any reason you wish to leave the testing room during an exam, you must first ask permission. If you happen to finish your exam(s) early and wish to leave the testing site before your designated session appointment is completed, you are permitted to do so only during specified dismissal periods.

UNDERSTANDING HOW YOUR EXAM IS SCORED

You can gain a better perspective about the ASE certification exams if you understand how they are scored. ASE exams are scored by an independent organization having no vested interest in ASE or in the automotive industry. With CBT exams, you will receive your exam scores immediately.

Each question carries the same weight as any other question. For example, if there are 50 questions, each is worth 2 percent of the total score.

Your exam results can tell you

- ■ Where your knowledge equals or exceeds that needed for competent performance, or
- ■ Where you might need more preparation.

Your ASE exam score report is divided into content "task" areas; it will show the number of questions in each content area and how many of your answers were correct. These numbers provide information about your performance in each area of the exam. However, because there may be a different number of questions in each content area of the exam, a high percentage of correct answers in an area with few questions may not offset a low percentage in an area with many questions.

It should be noted that one does not "fail" an ASE exam. The technician who does not pass is simply told "More Preparation Needed." Though large differences in percentages may indicate problem areas, it is important to consider how many questions were asked in each area. Since each exam evaluates all phases of the work involved in a service specialty, you should be prepared in each area. A low score in one area could keep you from passing an entire exam. If you do not pass the exam, you may take it again at any time it is scheduled to be administered.

There is no such thing as average. You cannot determine your overall exam score by adding the percentages given for each task area and dividing by the number of areas. It does not work that way because there generally is not the same number of questions in each task area. A task area with 20 questions, for example, counts more toward your total score than a task area with 10 questions.

Your exam report should give you a good picture of your results and a better understanding of your strengths and areas needing improvement for each task area.

Types of Questions on an ASE Exam

Understanding not only what content areas will be assessed during your exam, but how you can expect exam questions to be presented will enable you to gain the confidence you need to successfully pass an ASE certification exam. The following examples will help you recognize the types of question styles used in ASE exams and assist you in avoiding common errors when answering them.

Most initial certification tests are made up of 40 to 80 multiple-choice questions. The five-year recertification exams will cover the same content as the initial exam; however, the actual number of questions for each content area will be reduced by approximately one-half. Refer to Section 4 of this book for specific details regarding the number of questions to expect during the initial Electronic Diesel Engine Diagnosis Specialist (L2) certification exam.

Multiple-choice questions are an efficient way to test knowledge. To correctly answer them, you must consider each answer choice as a possibility, and then choose the answer choice that *best* addresses the question. To do this, read each word of the question carefully. Do not assume you know what the question is asking until you have finished reading the entire question.

About 10 percent of the questions on an actual ASE exam will reference an illustration. These drawings contain the information needed to correctly answer the question. The illustration should be studied carefully before attempting to answer the question. When the illustration is showing a system in detail, look over the system and try to figure out how the system works before you look at the question and the possible answers. This approach will ensure that you do not answer the question based upon false assumptions or partial data, but instead have reviewed the entire scenario being presented.

MULTIPLE-CHOICE/DIRECT QUESTIONS

The most common type of question used on an ASE exam is the direct multiple-choice style question. This type of question contains an introductory statement, called a stem, followed by four options: three incorrect answers, called distracters, and one correct answer, the key.

When the questions are written, the point is to make the distracters plausible to draw an inexperienced technician to inadvertently select one of them. This type of question gives a clear indication of the technician's knowledge.

Here is an example of a direct-style question:

1. The ECM has set a fault code for the camshaft position sensor. During troubleshooting the technician disconnects the camshaft position sensor wiring harness at the sensor and finds the internal terminals covered in engine oil. Which of the following could be the cause?

TASK B.9

   A. High engine oil pressure

   B. High engine crankcase pressure

   C. Failed camshaft position sensor

   D. Leaking valve cover gasket

Answer A is incorrect. The camshaft position sensor is sealed; oil pressure should not be able to enter the sensor. The sensor has failed.

Answer B is incorrect. The camshaft position sensor is a sealed sensor designed to operate in the presence of oil. If the sensor has oil covering the internal terminals as described in the question, the sensor has failed. High engine crankcase pressure is not indicated.

Answer C is correct. The camshaft position sensor has failed. It is no longer sealed and is allowing oil to enter.

Answer D is incorrect. A leaking valve cover gasket could allow oil to be on the outside of the sensor; however, the sensor connector is sealed and should not allow oil into the terminal area.

COMPLETION QUESTIONS

A completion question is similar to the direct question except the statement may be completed by any one of the four options to form a complete sentence. Here's an example of a completion question:

1. The coolant level sensor would be mounted in the:

TASK B.6

   A. Lower radiator tank.

   B. Cylinder head.

   C. Engine block.

   D. Surge tank.

Answer A is incorrect. The coolant level sensor needs to be mounted high in the cooling system. It would not be located in the lower radiator tank. The electronic control module (ECM) can be programmed to alert the driver then shut the engine off if the coolant level is low to prevent damage to the engine.

Answer B is incorrect. The cylinder head is higher in the cooling system than some of the other possible answers listed here; however, if the coolant level sensor was mounted in the cylinder head there would still be a significant chance of engine damage prior to the sensor alerting the driver of the low coolant level condition.

Answer C is incorrect. The coolant level sensor needs to be mounted high enough in the cooling system to alert the driver of the low coolant level condition before engine damage can occur. Mounting the sensor in the engine block would not provide this protection.

Answer D is correct. The coolant level sensor is mounted in the surge tank; the surge tank is one of the highest mounted components in the cooling system.

TECHNICIAN A, TECHNICIAN B QUESTIONS

This type of question is usually associated with an ASE exam. It is, in fact, two true-false statements grouped together, such as: "Technician A says…" and "Technician B says…", followed by "Who is correct?"

In this type of question, you must determine whether either, both, or neither of the statements is correct. To answer this type of question correctly, you must carefully read each technician's statement and judge it on its own merit.

Sometimes this type of question begins with a statement about some analysis or repair procedure. This statement provides the setup or background information required to understand the conditions about which Technician A and Technician B are talking, followed by two statements about the cause of the concern, proper inspection, identification, or repair choices.

Analyzing this type of question is a little easier than the other types because there are only two ideas to consider, although there are still four choices for an answer.

Again, Technician A, Technician B questions are really double true-or-false statements. The best way to analyze this type of question is to consider each technician's statement separately. Ask yourself, "Is A true or false? Is B true or false?" Once you have completed an individual evaluation of each statement, you will have successfully determined the correct answer choice for the question, "Who is correct?"

An important point to remember is that an ASE Technician A, Technician B question will never have Technician A and Technician B directly disagreeing with each other. That is why you must evaluate each statement independently.

An example of a Technician A/Technician B-style question looks like this:

TASK B.3

1. Technician A says incorrect calibration files in the ECM can cause poor fuel economy. Technician B says incorrect calibration files in the ECM can cause poor engine performance.

   A.   A only
   B.   B only
   C.   Both A and B
   D.   Neither A nor B

Answer A is incorrect. Technician B is also correct.

Answer B is incorrect. Technician A is also correct.

Answer C is correct. Both Technicians are correct. Incorrect calibration files can affect fuel delivery, idle quality, and power. When troubleshooting a diesel engine concern the technician should check to make sure the correct and latest calibration file is loaded into the ECM.

Answer D is incorrect. Both Technicians are correct.

EXCEPT QUESTIONS

Another type of question used on ASE exams contains answer choices that are all correct except for one. To help easily identify this type of question, whenever they are presented in an exam, the word "EXCEPT" will always be displayed in capital letters. Furthermore, a cautionary statement will alert you to the fact that the next question is different from the ones otherwise found in the exam. With the EXCEPT type of question, only one *incorrect* choice will actually be listed among the options, and that incorrect choice will be the key to the question. That is, the incorrect statement is counted as the correct answer for that question.

Be careful to read these question types slowly and thoroughly; otherwise, you may overlook what the question is actually asking and answer the question by selecting the first correct statement.

An example of this type of question would appear as follows:

1. A diesel engine that is not equipped with a diesel particulate filter (DPF) cranks but will not start. There is no smoke from the exhaust while cranking. All of the following could be the cause EXCEPT:

    A. No voltage to the injectors
    B. No fuel in the tank
    C. Low compression
    D. Failed engine position sensor

TASK B.4

Answer A is incorrect. No voltage to the injectors would prevent fuel injection; there would be no smoke and no start.

Answer B is incorrect. No fuel in the tank would mean no fuel injection; therefore, no smoke and no start.

Answer C is correct. Low compression can cause the engine not to start; however, there will be smoke because fuel will be delivered to the cylinders.

Answer D is incorrect. A failed engine position sensor can cause no fuel injection; therefore, no smoke and no start.

LEAST LIKELY QUESTIONS

LEAST LIKELY questions are similar to EXCEPT questions. Look for the answer choice that would be the LEAST LIKELY cause (most incorrect) for the described situation. To help easily identify these types of questions, whenever they are presented in an exam the words "LEAST LIKELY" will always be displayed in capital letters. In addition, you will be alerted before a LEAST LIKELY question is posed. Read the entire question carefully before choosing your answer.

An example of this type of question is shown here:

1. Which of the following would LEAST LIKELY be used to diagnose a misfiring cylinder on an electronic unit injector (EUI) fuel system?

    A. Scan tool
    B. Ohmmeter
    C. 1000ml beaker
    D. Pyrometer

TASK D.7

Answer A is incorrect. A scan tool can be used to run a cylinder contribution test on a diesel engine. This is a useful test in diagnosing a misfiring cylinder.

Answer B is incorrect. An ohmmeter can be used to measure the resistance of the solenoid on an EUI injector.

Answer C is correct. A 1000ml beaker is used to measure the drain system flow on an HPCR fuel system. It is not used on a EUI fuel system to diagnose a misfiring cylinder.

Answer D is incorrect. A pyrometer can be used to measure exhaust temperature at each cylinder to help diagnose a misfiring cylinder.

COMPOSITE VEHICLE QUESTION

Approximately 50 percent of the L2 certification exam questions refer to the Medium/Heavy Duty Composite Vehicle Type 3 Reference Booklet. The composite vehicle represents an imagined vehicle that contains diesel engine components contained in actual medium/heavy duty vehicles. You will need to refer to the reference booklet to answer questions about the composite vehicle. These questions are gathered together at the end of the section containing the more general questions relating to the ASE Task List for L2. Composite vehicle question follow the same style patterns as do the general questions. A sample composite vehicle question reads as follows:

TASK B.6

1. Refer to the composite vehicle to answer this question. Which of the following would be the most likely cause of the 15 amp vehicle circuit breaker continuously cycling open and closed?

   A. Open on circuit 178

   B. Short to ground on circuit 192

   C. Open on circuit 192

   D. Short to ground on circuit 178

Answer A is incorrect. An open on circuit 178 would cause no current flow; excessive current flow is what causes a circuit breaker to continually open and close.

Answer B is incorrect. A short on circuit 192 would cause excessive current flow; however, it would be before the circuit breaker and therefore would not affect the circuit breaker.

Answer C is incorrect. An open on circuit 192 would cause no current flow. The circuit breaker would not receive any voltage.

Answer D is correct. A short to ground on circuit 178 would cause increased current flow and would cause the circuit breaker to continually cycle.

SUMMARY

The question styles outlined in this section are the only ones you will encounter on any ASE certification exam. ASE does not use any other types of question styles, such as fill-in-the-blank, true/false, word-matching, or essay. ASE also will not require you to draw diagrams or sketches to support any of your answer selections, although any of the described question styles may include illustrations, charts, or schematics to clarify a question. If a formula or chart is required to answer a question, it will be provided for you.

Task List Overview

INTRODUCTION

This section of the book outlines the content areas or *task list* for this specific certification exam, along with a written overview of the content covered in the exam.

The task list describes the actual knowledge and skills necessary for a technician to successfully perform the work associated with each skill area. This task list is the fundamental guideline you should use to understand what areas you can expect to be tested on, as well as how each individual area is weighted to include the approximate number of questions you can expect to be given for that area during the ASE certification exam. It is important to note that the number of exam questions for a particular area is to be used as a guideline only. ASE advises that the questions on the exam may not equal the number specifically listed on the task list. The task lists are specifically designed to tell you what ASE expects you to know how to do and to help prepare you to be tested.

Similar to the role this task list will play in regard to the actual ASE exam, Delmar, Cengage Learning has developed six preparation exams, located in Section 5 of this book, using this task list as a guide. It is important to note that although both ASE and Delmar, Cengage Learning use the same task list as a guideline for creating these test questions, none of the test questions you will see in this book will be found in the actual, live ASE exams. This is true for any test preparatory material you use. Real exam questions are *only* visible during the actual ASE exams.

Task List at a Glance

The Electronic and Diesel Engine Diagnosis Specialist Test (L2) task list focuses on four core areas, and you can expect to be asked a total of approximately 45 questions on your certification exam, broken out as outlined:

    A.   General Inspection and Diagnosis (4 questions)

    B.   Electronic Engine Controls Diagnosis (27 questions)

    C.   Air Induction and Exhaust Systems Diagnosis (6 questions)

    D.   Fuel Systems Diagnosis (8 questions)

The L2 certification and recertification tests both cover the same content areas and have the same number of scored questions.

> *Note:* The L2 test could have up to 15 additional questions that are included for statistical research purposes only. Your answers to these questions will not affect your test score, but because you do not know which ones they are, you should answer all questions in the test.

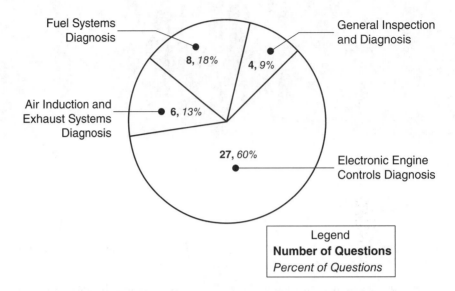

Legend
Number of Questions
Percent of Questions

Note: The actual number of questions you will be given on the ASE certification exam may vary slightly from the information provided in the task list, as exams may contain questions that are included for statistical research purposes only. Do not forget that your answers to these research questions will not affect your score.

TASK LIST AND OVERVIEW

A. General Inspection and Diagnosis (4 questions)

1. Identify engine model and serial number to research applicable vehicle and service information, service precautions, technical service bulletins, and service campaigns/updates; determine needed actions.

Engine model and serial numbers can be found on the engine data tag. This data tag may be located on the valve cover or accessory drive cover. In many cases it is also cast or stamped into the engine block. These numbers are used to locate relevant service information, service precautions, technical service bulletins, and service campaigns or updates. The model number and serial number found on the engine should match the numbers stored in the engine control module (ECM). If the numbers do not match, this could result in poor drivability, high exhaust temperatures, no-start, or low-power concerns. To restore the correct numbers in the ECM, it must be re-flashed with the proper information. This re-flash process can be performed with the ECM on the vehicle.

2. Verify operational complaint.

The first step in diagnosing a customer concern is to verify the operational complaint. If the customer concern cannot be duplicated, it may be necessary to ride with the driver to duplicate it. In some cases, customers may interpret a normal operational characteristic of the engine as a problem. An example would be a slightly elevated engine operating temperature. If the engine is performing a regeneration of the diesel particulate filter, a slightly elevated engine operating temperature is normal.

3. Determine if problem is electrical/electronic or engine mechanical.

A problem with a diesel engine can fall into the categories of electrical/electronic or engine mechanical. Possible electrical/electronic problems are opens, shorts, high resistance, and incorrect programming in the ECM. Examples of engine mechanical problems are broken valve springs, worn rings, worn camshaft lobes, and worn injector plungers. Determining the source of the problem will aid in correct diagnosis. For example, an engine that is hard to start could have low compression. Performing a mechanical cranking compression test will determine whether low compression is likely. An engine could also be hard to start due to low cranking speed. Low cranking speed can be caused by high resistance in the battery cables. A voltage drop test on the battery cables while the engine is being cranked will determine whether the cables have high resistance.

4. Evaluate engine mechanical condition based on a visual inspection of the exhaust output. (Applies to engines with a damaged diesel particulate filter [DPF] and to engines not equipped with a DPF.)

Exhaust smoke output contains valuable diagnostic information for engines that are not equipped with a DPF. White smoke can be the result of incorrect injection timing, low fuel supply volume/pressure, or low compression. Black smoke can be caused by excessive fuel delivery or insufficient air delivery. Blue smoke can be caused by oil entering the combustion chamber. Smoke in an engine with a DPF is an indication of a damaged DPF. The diesel particulate filter can be damaged by engine oil or coolant contamination, overfueling the engine, or a failed after treatment injector.

5. Check and record electronic diagnostic codes, freeze frame and/or operational data; interpret live engine data; download/save ECM data (image); determine further diagnosis.

During a general inspection and diagnosis the technician should retrieve and store the trouble codes. These trouble codes should be recorded on the work order. If the code is currently inactive, the technician should note the first occurrence, most recent occurrence, and the number of occurrences. A high number of occurrences may indicate a loose connection or damaged wiring harness. Active codes should be diagnosed using the troubleshooting trees supplied by the manufacturer. Live engine data can be looked at by the technician to help locate sensors that are operating within operational parameters, but not within normal conditions. For example, if the truck has set 24 hours without being run and the air inlet sensor indicates 120°F while the outside air temperature is 65°F there is very likely a problem with the sensor. However, because 120°F would be considered a temperature that is possible, the sensor reading would not set a DTC. Engine data can be stored in a separate file and saved to a hard drive and/or attached electronically to a work order to have a complete history of the diagnosis and repair.

6. Diagnose performance complaints caused by engine cooling system problems.

Overheating, overcooling, low power, and engine shutdown can all be complaints caused by the engine cooling system. Diagnosis of cooling system complaints should start with verifying that the correct amount and type of coolant is installed in the engine. Coolant condition can be checked with a refractometer and/or coolant test strips. Cooling system pressure can be checked with a cooling system pressure tester. Causes of excess cooling system pressure include blown head gasket or cracked head. External coolant leaks can be caused by loose mounting bolts, seeping gaskets or o-rings, warped gasket surfaces, or cracked components. Sources of internal coolant leaks are jacket water aftercoolers (JWAC), exhaust gas recirculation coolers (EGR), liner o-rings, cracked cylinder heads, gear-driven water pumps, or blown head gaskets.

7. Diagnose performance complaints caused by engine lubrication system problems.

Engine lubrication system failure can damage bearings, pistons, cylinder walls, and other internal components. Performance complaints associated with lubrication system failures include no-start, start and dies, engine noise, and particulate filter restrictions. To diagnose lubrication system failures, the technician should begin by checking the level and quality of the oil in the oil pan; the major cause of low oil pressure is low oil level. Some engines are programmed to shut down when the oil level or oil pressure falls below a prescribed threshold. These conditions will usually be accompanied by a diagnostic trouble code (DTC). If low oil pressure is suspected, the oil pressure should be verified by connecting a master gauge to the engine. Because oil pressure sensors and gauges installed on trucks are prone to failure, many low oil pressure concerns result from failures in the oil pressure sending unit. Using incorrect grade or weight engine oil can cause the diesel particulate filter to regenerate more often than normal. Because the technician cannot field test the oil to determine if it is correct, service records must be relied upon. Additionally, the technician may use the scan tool for diagnostics to determine how often the ECM is cleaning the DPF.

8. Evaluate the integrity of the air induction system.

Air induction systems include all the components from the air filter inlet to the intake valve. The integrity of the air inlet system is critical to the longevity of the engine. The system should pass a thorough visual inspection, paying particular attention to the areas where intake tubing is mounted and where other components may rub against it. Often, a hose clamp or hose laid across an air intake pipe will slowly wear a hole in the pipe. This may be hard to detect because the item that made the hole is still lying on top of the hole. It is important for the technician to move components around during the inspection to look for these conditions. Traces of dust on the inside of air intake piping are a good indication that the system is leaking. To check the charge air cooler for leaks, the technician should pressurize the cooler to approximately 25 psi. Close off the air supply to the tester and time how long it takes for the cooler to leak off the pressure. A typical specification is a maximum leak rate of 5 psi in 15 seconds.

9. Evaluate the integrity of the exhaust system.

The exhaust system must be free from leaks and securely mounted. As of 2007, exhaust systems can no longer utilize flex piping. By its very nature, flex piping will leak. If the piping is installed before the DPF, leaking exhaust will contain higher than legal limits of emissions. On late-model trucks, the exhaust connections with V-band clamps have gaskets installed. The gaskets provide a leak-free joint to ensure the engine does not leak untreated exhaust. These gaskets should be replaced anytime the connection is open. The DPF should be mounted correctly and all wiring harnesses tied up securely. The exhaust system cannot be modified from its original configuration without manufacturer and EPA approval. Any reconfigurations made without approval can increase exhaust emissions and are therefore illegal.

10. Listen for and isolate engine noises; determine needed actions.

Engine noises can be located using a stethoscope or similar device. Noises can be isolated by using a scan tool to run a cylinder contribution test. If the noise being generated changes when a particular cylinder is cut out (cancelled), it is a good indication that something associated with that cylinder is the cause. Noise can be caused by any of the following: worn or misadjusted valves or injectors, faulty fuel injectors, collapsed pistons, worn piston pins, worn main or rod bearings, loose flywheels, damaged harmonic balancers (vibration dampers), and gear trains with excessive or insufficient backlash. Noises that come and go with engine RPM are typically harmonic balancer problems. The balancer (damper) can be measured with a dial indicator and/or a micrometer to help identify failed components.

11. Diagnose performance complaints caused by drive train and tire problems or modifications.

Performance complaints that originate from drive train and tire problems or modifications are often tied to incorrect speedometer readings, vehicles not reaching maximum speed, and low fuel economy. If the vehicle tire size and/or RPM are incorrectly entered into the engine ECM, the engine ECM will incorrectly regulate maximum vehicle speed and incorrectly calculate fuel economy. If the speedometer receives its signal from the ECM or body control module (BCM) over the data bus, this incorrect data will cause the speedometer and odometer to read incorrectly. If the installed vehicle ring gear and pinion have a different gear ratio than those programmed in the ECM, the same conditions will occur.

12. Diagnose performance complaints caused by vehicle operation and/or configuration of mechanical and electronic components.

Technicians should consider vehicle operation when diagnosing a performance concern. A DPF that needs unusually frequent cleaning of accumulated soot may be due to the way the vehicle is being operated rather than a component failure. Vehicles that are lightly loaded and operate at primarily slow speeds may not reach the required operating conditions for the ECM to perform a regeneration of the DPF. If the ECM cannot perform the regeneration, it will be necessary for the technician to perform active stationary regeneration to clean the filter. This may be a normal condition on some vehicles. It would not, however, be a normal condition for a vehicle that is in a line haul operation where sustained speeds and temperatures allow the regeneration to occur during the normal driving cycle.

13. Diagnose no-crank, cranks-but-fails-to-start, extended cranking, and starts-then-stalls; determine needed actions.

No-crank conditions can be caused by low batteries, open battery cables, and failed starters. Cranks-but-fails-to-start can be caused by low fuel level, restricted fuel systems, failed ECMs, blown fuses, a failed fuel shutoff device, and extremely restricted exhaust systems. Extended cranking is typically caused by fuel drain-back problems. Starts-then-stalls can be caused by air in the fuel system, low fuel level, low coolant level, and/or low oil level. Electrical problems can be diagnosed using voltmeters, ohmmeters, and ammeters. Battery condition can be checked using a carbon load tester or a battery capacitance tester. Fuel system problems are typically diagnosed with a fuel pressure gauge or flow test.

14. Visually inspect engine compartment wiring harnesses and connectors; check for proper routing, condition, and mounting hardware; determine needed actions.

Every diagnosis should include a thorough under-the-hood inspection. Many problems associated with electronic diesel engines stem from worn or damaged electrical connectors and wiring harnesses. Connectors should be checked for loose pins, missing connector locks, and loose sockets. All wiring harnesses should be securely mounted away from heat and moving components. Mounting hardware should be properly tightened and correctly positioned to prevent undue strain on the harness or electrical connectors. Some manufacturers approve the replacement of pins and sockets within electrical connectors. Others provide a new electrical connector with a short pigtail to crimp onto the original wiring harness.

15. Diagnose surging, rough operation, misfiring, low power, slow acceleration, slow deceleration, and shutdown problems; determine needed actions.

Surging can be caused by air in the fuel or faulty electronic signals. Rough operation can be caused by worn or damaged injectors, air in the fuel, or a cylinder with low compression. Slow acceleration and deceleration can be caused by restricted fuel lines, restricted exhaust or intake, or air in the fuel. Shut-down problems are typically the result of an insufficient fuel supply or programming within the ECM, which shuts the engine down for protection, as in the case of engine overheating, low coolant, or low oil pressure. Diagnosing the concerns and determining the needed actions are typically performed using scan tools, pressure and vacuum gauges, and thorough visual inspections.

16. Determine root cause of current, multiple, and/or repeat failures.

Usually when a truck has repeat failures the root cause has not been addressed. For example, a truck that has blown multiple head gaskets may have a warped cylinder block or head, stretched head bolts, incorrect turbo boost, or faulty ECM programming. When a technician is repairing these vehicles, it is critical that the root cause be found to prevent the need for repeat repairs.

17. Verify effectiveness of repairs and clear diagnostic codes (if applicable).

After repairs are completed, the technician must road test or otherwise operate the vehicle to ensure the repairs have fixed the original concern.

Concerns that involve heavy-duty on-board diagnostic codes (HD OBD) may not immediately extinguish the malfunction indicator lamp (MIL) after the repair. Starting in 2010, engines that show HD OBDs may require multiple trips to clear the MIL. It is possible that the technician could repair a vehicle correctly and operate it, yet the MIL would remain on while the stop engine lamp (SEL) and check engine lamp (CEL) lamp are off. Some manufacturer's scan tools will allow the technician to reset the MIL as well as remove the fault code.

B. Electronic Engine Controls Diagnosis (27 questions)

1. Inspect and test for missing, modified, or damaged engine control components.

It is illegal to remove, tamper with, or modify any original equipment emission control systems. During diagnosis a technician needs to be aware that some components may be missing or modified. Exhaust gas recirculation (EGR) valves and DPFs will sometimes be removed in an attempt to improve fuel economy or power. Occasionally, the ECM will have been re-flashed by the individual who removed these items in order to prevent DTCs from setting.

A thorough visual inspection will often provide evidence of block off plates and/or disconnected or damaged wiring harnesses. To test, the technician can use a scan tool to command the ECM to operate devices like the EGR valve or force the engine into a stationary DPF regeneration. If components are missing, damaged, or modified, they must be replaced or repaired to match the original equipment.

2. Check and record electronic diagnostic codes, freeze frame and/or operational data; interpret live engine data; download/save ECM data (image); determine further diagnosis.

Electronic diagnostic codes can be retrieved using a scan tool or an on-board display. All codes, whether active or inactive, should be recorded on the shop work order. Active codes can be diagnosed using the appropriate service literature. Inactive codes cannot be diagnosed as though they are currently active; however, the technician will sometimes encounter high counts of inactive codes. These can be an indication of loose connections or other intermittent conditions.

Most engine ECMs store freeze frame data at the time a DTC is set. This data stores the readings of sensors, engine parameters, ECM time, and the operating condition of most engine actuators. This data can tell the technician important information such as the RPM and temperature of the engine when the failure occurred (ECM time). Live engine data can be checked with a scan tool by the technician to determine all the information that the ECM is receiving from the sensors as well as the commands issued by the ECM to actuators. Freeze frame data can be stored in a separate file by making an image of the ECM. This image can be stored with the work order to have a complete history of the diagnosis and repair.

3. Connect diagnostic tool to vehicle/engine; access, verify, and update parameters and calibration settings; perform updates as needed.

The ECM has several customer-adjustable (programmable) parameters. These parameters need to be checked to ensure they are set properly. For example, if the fleet is intending to help the driver improve fuel economy and use cruise control, the "maximum vehicle speed with cruise" will usually be set a few miles per hour faster than the "maximum vehicle speed without cruise" parameter. The ECM should always be checked for the latest calibration update settings. It is very possible that the manufacturer has issued an update calibration file that specifically addresses the concern on the work order. In this case, an updated calibration is the appropriate repair. Instructions on the scan tool should be followed exactly when performing updates. It is equally important that neither the scan tool nor the engine ECM lose power during the process. A power failure or equipment disconnect during the update process can ruin the engine ECM.

4. Determine if the control system problem is electrical/electronic or mechanical.

To determine if the control system problem is electrical/electronic or mechanical, the technician needs to follow a troubleshooting analysis tree. Through a series of steps, the technician will electrically test the wiring harness and control module. If the electrical/electronic tests pass, then the problem is mechanical. Mechanical faults could include a failed EGR cooler, blocked EGR passages, or restricted DPF. Electrical/electronic failures would include opens, shorts, and failed solenoids or relays.

5. Use a diagnostic tool to inspect and test the electronic engine control system, sensors, actuators, electronic control modules (ECMs), and circuits; determine further diagnosis.

A diagnostic tool can be used to test sensors by comparing the sensor values on the scan tool to actual measured values. For example, a technician can access the coolant temperature sensor data on the scan tool and then use an infrared thermometer to check actual engine temperature. These two values should be very close to one another. Another method is to compare sensors. If a truck has been sitting for 24 hours without being run, the sensor data values for engine coolant temperature and air inlet temperature should be nearly identical. If they are not, it may indicate a failed sensor, wiring harness, or ECM. Actuators are usually tested with the actuator test mode of the diagnostic tool. This mode will command the actuators to function when commanded by the technician. The injector buzz test is an example of an actuator test. If one injector failed to operate during the tests, the technician would continue the diagnosis by testing the wiring harness and possibly measuring the resistance of the injector solenoid.

6. Test and confirm operation of electrical/electronic circuits not displayed on diagnostic tools.

Many circuits that will influence the operation of a diesel engine are not displayed on the scan tool. Some examples are the starting and charging systems. The starting system can cause a no-start or hard-to-start condition due to weak batteries, high electrical resistance in the wires or cables, or a faulty starter motor. The system is checked using a voltmeter and battery tester. A faulty charging system can cause the engine to die or run rough. If the charging system is undercharging, the injectors may not receive sufficient voltage to operate properly, which could result in a rough-running engine. Additionally, the ECM may be programmed to shut the engine off if battery voltage drops below a predetermined level, as a safeguard against excessive current flow through the ECM. A charging system with an open diode can cause A/C voltage to pass through the vehicle's electrical system. This A/C voltage can cause erratic operation of the engine. The charging system can be checked using a voltmeter, an oscilloscope, and a load tester.

7. Diagnose engine problems resulting from failures of interrelated systems (for example: cruise control, security alarms/theft deterrent, transmission controls, electronic stability control, non-OEM-installed accessories).

Many electrical systems on a vehicle are interconnected, usually through the vehicle data bus. A failure in one system can cause another system to operate incorrectly. A customer may have a complaint of no engine brake operation. This can be caused by a failure of the engine brake circuit or the engine ECM, or it could be caused by the antilock brake system (ABS). When the ABS recognizes a sliding wheel end, indicated by a wheel speed sensor showing a wheel end slowing down faster than the others, it will send a command through the data bus to the engine ECM to disable the engine compression brake. Therefore, a wheel speed sensor sending a faulty signal, or an ABS controller misinterpreting that signal, may result in a seemingly totally unrelated failure. This type of failure may not set a trouble code because wheel end sliding is a normally occurring event, and therefore not recognized as a system fault. An effective troubleshooting method is to remove controllers from the data bus to determine if the concern disappears. Some engine ECMs can be temporarily set to ignore all incoming data bus messages in order to perform this type of troubleshooting.

8. Measure and interpret voltage, voltage drop, amperage, and resistance readings using a digital multi-meter (DMM) or appropriate test equipment.

Voltage checks help the technician determine available voltage. While the circuit can be opened to check available voltage, it is much more accurate to check available voltage to the load while it is connected and operating. This will indicate whether voltage is being dropped prior to the load. Voltage drop is measured using a voltmeter connected across the circuit with the circuit operating. Typically, voltage drops should not exceed 0.5 volts. A voltage drop test cannot be performed on a circuit that is not operating. Resistance measurements are taken with an ohmmeter. The circuit or device being tested must be de-energized and disconnected. Amperage can be tested by using an amp clamp or inserting an ammeter in series with the circuit to determine current flow. High circuit resistance can be caused by low current flow, but this can only be determined in an operational circuit.

9. Inspect and repair/replace electrical connector terminals, pins, harnesses, seals, and locks.

There are many styles of connectors and terminals. Some are push-to-seat, others are pull-to-seat, and in others the terminals are not removable and the entire connector must be replaced. Regardless of the style, the fit between electrical terminals and pins can loosen over time. The technician should perform a pin fit test to check for looseness by sliding a new connector into a used connector. The fit should be snug. If the fit is loose, the used connector should be replaced. Terminals should also be checked to ensure they fit securely in the body of the connector. A broken connector lock will allow the terminal to slide out the back of the connector body when the two halves of the connector are joined. Harnesses should be secure and away from rotating or moving parts. Seals should be in place, not hardened or ripped. A failed seal will allow moisture in the connector, resulting in a corroded connection and, eventually, a high-resistance circuit. Locks should be in place and secure. Often, locks are broken when the technician disconnects a harness connector for troubleshooting. If a lock is broken, it must be replaced to prevent circuit failure caused by the harness connector loosening during operation.

10. Determine root cause of current, multiple, and/or repeat failures.

Repeat failures almost always have a root cause that has not been corrected. In the engine electrical control system, a common repeat failure is a recurring DTC. For example, an engine may show a DTC for the variable geometry turbocharger (VGT). The technician could diagnose the concern as a failed VGT actuator, because when the technician tested the arm and the swinging vanes within the VGT, they moved freely. However, if the original problem occurred with the engine at operating temperatures, it may be that the swinging vanes begin to stick when they are hot. The VGT actuator could thus be replaced needlessly if the test is performed under different conditions.

11. Verify effectiveness of repairs and clear diagnostic codes (if applicable).

After repairs are made, the technician must check that the original problem no longer exists and no new problems have arisen. Typically, a test drive will help verify the effectiveness of the repair. During the test drive, the technician should operate the vehicle with various loads, making sure the operating conditions that were present when the original problem occurred are duplicated.

After the test drive, the technician should connect the scan tool and check for trouble codes as well as review the live data stream to make sure the sensor and actuator values are within normal operational parameters.

C. Air Induction and Exhaust Systems Diagnosis (6 questions)

1. Perform air intake system restriction, pressure, and leakage tests; determine needed actions.

Air intake restriction tests are performed using a water manometer or an electronic vacuum gauge that has been calibrated to read inches of water. The test needs to be performed with the engine under full load. The normal maximum air inlet restriction specification is 25 inches of water.

The intake system should be checked for pressure using a 0-75 psi pressure gauge. The pressure should be checked before and after the charge air cooler. Boost pressure specifications will vary considerably between engines, and the manufacturers' specifications should be verified. To perform the boost pressure check, the engine must be operated under full load.

The charge air cooler can be checked for leaks by using a charge air cooler test kit. This kit installs on the air inlet and outlet of the charge air cooler and allows the technician to fill the charge air cooler to approximately 25 psi. When the specified air pressure is reached in the cooler using shop air, the air supply is shut off. The technician measures how long it takes for the air pressure to leak. A normal specification is up to 5 psi drop in 15 seconds.

2. Inspect, test, and replace intake air temperature and pressure sensors.

Air intake temperature sensors must be checked for accuracy. This is easily performed by comparing the reading of the air intake temperature on the scan tool to the air intake temperature measured using an infrared temperature gun. The two readings should be very close. If the vehicle has been sitting overnight, the technician can connect the scan tool and compare the temperature reading to the ambient air temperature as well as to the temperature sensors on the engine. All the temperature sensors should report approximately the same temperature. If not, the sensor that is reporting a different temperature is suspect.

Pressure sensors should be checked key-on, engine-off using a scan tool. With the key on and engine off, the sensor should report 0.0 psi gauge (psig) or approximately 14.7 psi absolute (psia). If the sensor passes this test, the technician should apply a pressure of approximately 20 psi using a pressure/vacuum gun and compare this reading to the reading on the scan tool.

3. Inspect and test turbocharger(s) (including variable ratio/geometry VGT), pneumatic, hydraulic, electronic controls, actuators, and sensors. Inspect, test, and replace wastegate and wastegate controls.

Turbochargers should be inspected for leaks in the lubrication system connections as well as the exhaust and intake connections. The turbocharger shaft should be checked for axial and radial movement using a dial indicator. A spin test by hand, while pushing and pulling on the shaft, allows the technician to determine if there is enough play to allow the compressor or turbine wheel to contact the housing. If so, the turbocharger must be replaced. A more detailed check would include using a dial indicator and comparing actual movement to manufacturer's specifications.

Variable geometry turbochargers (VGTs) have a sliding ring or swinging vanes that control the exit area of the exhaust gases. These moving components are subject to wear and prone to sticking. The technician can use the scan tool to command the VGT actuator to move the ring or vanes through its entire movement. If the components are binding, the ECM will set a trouble code. Some manufacturers recommend removing the exhaust housing of the turbocharger and cleaning the VGT components. VGT actuators can be air, hydraulic, or electronically controlled. Air and hydraulic actuators must be checked for leaks and restrictions to flow. Electronic actuators can be simple actuators or smart devices that communicate over a private 1939 data bus.

Wastegates are mounted on the exhaust turbine housing. They allow some of the exhaust gas to bypass the exhaust turbine wheel of the turbocharger to control turbocharger boost pressure. A wastegate that is stuck closed can cause overboost, while a wastegate that is stuck open can cause underboost. Wastegate control systems can be mechanical or electronic. The control system should be checked for loose, leaking, or restricted hoses. Excessive turbocharger boost can lead to premature failure of the turbocharger and engine.

A failed turbocharger can allow large quantities of oil into the intake or exhaust system. If this has occurred, the technician will need to clean the charge air cooler and/or the diesel oxidation catalyst (DOC) and DPF.

4. Perform exhaust backpressure and temperature tests (if applicable); determine needed actions.

Exhaust backpressure checks are performed using a mercury manometer while the engine is under full load. On older engines, a typical specification of about 3 in. Hg was considered the maximum exhaust backpressure. Late-model engines with a DOC/DPF use pressure sensors in the exhaust system that allow the technician to measure the pressure with a scan tool. On DOC/DPF-equipped engines, exhaust backpressure specifications will vary widely and the appropriate service literature must be consulted. High exhaust backpressure is normally caused by exhaust restriction. A collapsed piping or muffler can restrict exhaust. A restricted DOC/DPF in the exhaust can be caused by an ECM that is not performing a regeneration when needed or by a large quantity of contaminants, such as fuel or oil, entering the exhaust. The DOC/DPF can be removed and cleaned in a machine specifically designed for that purpose.

Exhaust temperature is an indication of engine condition. Low exhaust temperatures can indicate that an engine is receiving insufficient fuel. High exhaust system temperatures can be due to restricted exhaust or engine overfueling. Engines with a DOC/DPF have temperature sensors mounted in the exhaust stream that the technician can monitor using the scan tool. During a DPF regeneration event, the technician can monitor these temperature sensors to help determine the condition of the DOC and DPF.

5. Inspect and test pre-heater/inlet air heater and/or glow plug system and controls.

Some diesel engines are equipped with an electrically operated grid style pre-heater mounted in the air inlet system. This pre-heater is turned on by the ECM when the engine is cold to aid in starting and help limit smoke during engine warm-up. Some engines also use a glow plug system to heat the combustion chambers to achieve the same effect. A few engines utilize both systems.

These heaters can be checked for proper operation by using an amp clamp to measure current flow while in operation. Current flow specifications will vary, and the specifications for the engine being tested should be checked. Typically, each glow plug will draw about 10 amps, and an air inlet heater will draw about 80 amps. These devices almost always fail open; therefore, a failed heater or glow plug will normally have no current draw. The technician can also use an ohmmeter to measure the resistance of the heating element. A third test, using a test light, requires connecting the alligator clamp of the test light to the battery positive terminal while connecting the tip of the test light to the positive terminal of the glow plug. If the test light illuminates then the glow plug is providing a ground and is operational.

The control system is usually controlled by the ECM. The ECM relies on the temperature sensors mounted on the engine to determine the appropriate on time. A failed or shifted temperature sensor can cause the system to stay on too long or not long enough. The glow plugs or air inlet heater can be commanded to the on position by using a scan tool to aid in testing.

6. Inspect and test the exhaust after treatment system; verify regeneration operation. Replace after treatment mechanical and electronic components as needed.

The exhaust after treatment system should be inspected for loose mounts and evidence of exhaust leaks. The wiring harness should be securely mounted away from exhaust components. The hoses should be tight and free from abrasions or cuts. The regeneration process can be tested by performing a stationary regeneration using the dash-mounted switches or the scan tool. When performing a stationary regeneration, the exhaust system and the exhaust exiting the tailpipe will be extremely hot, so all manufacturer's safety precautions must be followed exactly. Some after treatment systems are made to be separated and the individual sections replaced. These systems are typically assembled with V-band-style clamps and have seals between the components. The technician should always replace the seals when servicing this unit and torque the V-band clamp to the correct specification. When replacing electrical components, the wiring harness and connectors must be securely positioned away from heat; all torque specifications must be followed. After any repairs to the system, the truck should be taken on a test drive and checked with a scan tool to verify that the system is operating properly.

7. Inspect and test EGR system components, including EGR valve(s), cooler(s), piping, sensors, controls, and wiring.

The EGR valve can be tested with the scan tool. The technician will command the EGR valve to operate and observe the EGR valve position sensor to monitor the response to the command. EGR valves are prone to sticking and becoming restricted with carbon. Some manufacturers provide information on the proper method to clean the EGR valve and passages. When the EGR system fails, the technician may be able to clean the system and restore it to operation. Because the EGR cooler can leak exhaust or coolant, the cooler should be inspected for leaks. Typically, coolant leaks will allow coolant into the air stream. These leaks are internal and can usually be located by removing air intake components downstream from the cooler and inspecting. Leaking EGR coolers are normally replaced instead of repaired.

8. Inspect and test EGR airflow control (throttle) valve systems and controls.

Some EGR systems incorporate a butterfly valve in the air intake system. When the engine is operating under conditions that require EGR flow, the ECM can partially close this valve to restrict incoming fresh airflow. This restriction helps create a low pressure area, which enhances the flow of exhaust gases into the intake manifold. These systems can normally be commanded to operate with the scan tool while checking for proper response. During engine operation, the ECM relies on pressure sensors in the intake and exhaust systems to determine when this butterfly valve needs to be operated.

9. Inspect and test variable valve actuator systems and controls.

Some manufacturers use variable valve actuators to help control the operation of engine intake valves, thus enhancing airflow in the combustion chamber and reducing emissions. These actuators use engine oil pressure to alter the timing of intake valves under certain operating conditions. The scan tool can be used to verify proper variable valve actuator operation.

10. Inspect and test crankcase ventilation system components.

The EPA now includes crankcase emissions from the crankcase ventilation system as part of the total emissions generated from the engine. Most manufacturers now have a crankcase filter to separate the oil vapor from the air exiting the crankcase through the ventilation tube. These oil vapors are returned to the crankcase while the air exits the ventilation tube. The system has a crankcase pressure sensor that monitors pressure to indicate a restricted filter. It should be noted that using this sensor is not a recommended method of measuring crankcase pressure as an aid in diagnosing a worn engine. The crankcase pressure ventilation system can be tested with a scan tool. The readings obtained from the sensor are then compared to the crankcase pressure readings measured while using a gauge.

11. Determine root cause of current, multiple, and/or repeat failures.

The root causes of air induction system failures can often be traced to incorrect fuel delivery and, therefore, require a thorough examination of the fuel delivery system. Low power, low turbocharger boost pressure, and smoke are all failures that can result from low fuel delivery volume and/or pressure. Over-fueling and/or a stuck open injector can cause smoke, high exhaust temperatures, and damaged DOC/DPF. Additionally, a failed turbocharger passing oil into the exhaust stream is a common cause of DOC/DPF face-plugging failures.

12. Verify effectiveness of repairs and clear diagnostic codes (if applicable).

After repairing the air intake and exhaust systems, the technician should road test the vehicle to verify that the original concern was repaired and that no new concerns have arisen after the repair. If possible, this road test needs to be performed while operating the vehicle under load. Any new fault codes arising during the test drive need to be diagnosed and repaired before returning the vehicle to the customer.

D. Fuel Systems Diagnosis (8 questions)

1. Determine if the fuel control system problem is electrical/electronic or mechanical.

The fuel system should be leak free and filled with high-quality, fresh diesel fuel. Since 2007, all on-highway diesel fuel is required to be ultra-low-sulfur diesel fuel (USLD). USLD has fewer than 15 parts per million (PPM) of sulfur. Using fuel that does not have this rating is illegal and will result in early DPF failures. To determine whether the fuel system problem is mechanical or electronic, the technician can perform an injector click test. If the injector does not click during this test, the problem is electronic. If the injector does click, the problem is mechanical. It could be mechanical within the injector or it could be mechanical within the cylinder, as in the case of a cylinder with low compression.

2. Check fuel system for air; determine needed repairs; prime and bleed fuel system.

The fuel system can be checked for air in a variety of ways. The technician can install a liquid eye or a clear piece of tubing on the suction side of the fuel system prior to the fuel transfer pump. Operating the engine with this installed will allow the technician to see whether air is traveling with the incoming fuel supply. Submerging the return fuel line in a container of fuel, starting the engine, and watching the return fuel stream for the presence of bubbles is another useful test. It should be noted that if bubbles are present in this latter test, they can be due to suction-side fuel system leaks or to faulty injectors that allow combustion chamber gases to enter the fuel system.

3. Inspect and test fuel supply system pressure, restriction, and return fuel rates; check fuel for contamination; determine needed repairs.

Fuel supply system pressure and flow rates should be checked with the engine loaded. The fuel system may be able to supply sufficient fuel pressure/volume when the engine is operated at idle or with a light load, but unable to keep up with demand when the fuel quantity requirements are high, such as when the engine is under full load. To aid in testing for this condition, some manufacturers have developed tools that incorporate a calibrated orifice to simulate the engine operating under load. When using these tools, the technician must follow the directions exactly. Failure to use the adapter with the correct restriction orifice in the proper location, or failure to perform the test for the specified amount of time, will provide inaccurate results and result in an incorrect diagnosis. Fuel contamination can be checked by pulling a fuel sample from the vehicle fuel tank or the fuel supply system. The fuel sample may need to sit for a few minutes, thereby allowing contaminants to settle out so that a correct interpretation may be rendered. Pastes are available that can be put on a rod and inserted in the fuel to check for the presence of water in the fuel tank. The paste will change color if water is present in the fuel. Another concern is when a failed fuel system allows engine oil to enter the fuel. Dark fuel would indicate this condition.

Fuel return rates should be checked on some fuel systems. To check fuel return rates, place the fuel return line in a graduated container and run the engine at the correct RPM for the specified amount of time. Low fuel return rates indicate a restricted supply system. Excessive fuel return rates can indicate failed injectors on some fuel systems.

4. Inspect, adjust; repair/replace electronic throttle and power take-off (PTO) control components, circuits, and sensors.

Accelerator pedal position (APP) sensors are typically potentiometer or Hall Effect types of sensors. They can send a single signal or multiple signals to the ECM. Multiple signal styles use the extra signals as a cross-reference safety measure. Potentiometer-type sensors can be tested with a scan tool and/or an ohmmeter. Hall Effect sensors are typically tested using the scan tool. Failed electronic throttle sensors and circuits will normally only allow the engine to idle.

PTO speed control circuits are often switches that send a signal to the ECM. The ECM raises engine RPM to the value set in the adjustable parameters menu. System access to change PTO set speed may require a password from the manufacturer.

5. Inspect, test, and replace high-pressure common rail (HPCR) fuel system electronic and mechanical components.

High-pressure common rail fuel systems (HPCR) use a mechanically driven pump to create fuel pressure in excess of 20,000 psi. These high-pressure pumps typically have two or three plungers and are of the inlet metering design. The ECM senses the fuel pressure in the fuel rail and adjusts the fuel coming into the high-pressure pump to control fuel pressure in the rail. To diagnose the system, the technician can monitor desired fuel pressure, actual fuel pressure, and the signal command to the fuel inlet metering valve using the scan tool. A high-pressure relief valve on the common rail releases system pressure in case of over-pressurization. The ECM energizes the injector electrically to allow fuel to flow through the injector body and into the combustion chamber. Although the fuel lines connecting the HPCR to the injector look similar to the injector lines used on older pump-line-nozzle (PLN) fuel systems, it is not possible to crack a line to help isolate a misfiring cylinder. All cylinder deactivation on this system must be done using the scan tool. When replacing components on this system, cleanliness is extremely important. Since common rail fuel injectors are solenoid-operated and do not require a camshaft and/or rocker to create high pressure, they do not require adjustment after installation.

6. Inspect, test, and replace hydraulic electronic unit injection (HEUI) fuel system electronic and mechanical components.

The HEUI fuel system uses high-pressure oil to activate the fuel injector. The high-pressure oil is generated by a high-pressure oil pump that receives its oil from the engine lubrication oil pump. The ECM controls the high-pressure oil pump through a pulse-width modulated (PWM) valve. Using a scan tool, the technician can monitor desired high-pressure oil pressure, actual high-pressure oil pressure, and the pulse width or control circuit current on the PWM control valve. The fuel pump on a HEUI fuel system delivers diesel fuel to the injector at approximately 100 psi. Oil mixing with the fuel on this fuel system is not uncommon and is typically caused by failed injectors or failed injector o-rings. Using the scan tool, injectors can be tested with an injector buzz test or a click test. When conducting a high-pressure oil pump system test on the oil control circuit, the pump is commanded to operate at multiple high-pressure settings while the pulse width of the control valve needed to maintain the high pressure is monitored.

When replacing HEUI injectors and some other styles of injectors, it is sometimes recommended to replace the hold-down bolts with new ones. The service information should always be followed. HEUI injectors do not require adjustment after installation.

7. Inspect, test, and replace electronic unit injection (EUI) fuel system electronic and mechanical components.

Electronic unit injector (EUI) fuel systems use a camshaft-operated unit injector to create the high pressure needed for injection. Fuel is supplied to the injector from a mechanical fuel transfer pump at about 80 psi. The ECM controls the injector by supplying voltage to the solenoid. The solenoid controls fuel flow in the injector. The injector's solenoids can be tested using a scan tool or an ohmmeter. The scan tool can be used to perform a cylinder power balance test. When replacing the fuel injector, it will need to be adjusted. Some manufacturers require a special height tool to perform this adjustment.

8. Determine root cause of current, multiple, and/or repeat failures.

Multiple or repeat failures of the fuel system are often caused by poor-quality fuel. Clean fuel is critical to component life in the fuel system. When repeat failures occur, the technician should check the fuel supply system for incorrect, failed, or incorrectly installed fuel filters. If a fleet is experiencing repeat failures on multiple vehicles, the fleet fuel supply system should be checked for contaminants. The fuel handling procedures of the fleet should also be reviewed.

9. Verify effectiveness of repairs and clear diagnostic codes (if applicable).

After repairs are made to the fuel system, the technician must confirm that the original problem no longer exists and no new problems have arisen. Typically, a test drive will verify the effectiveness of the repair. During the test drive, the technician should operate the vehicle with various loads, ensuring that the operating conditions present when the original problem occurred have been duplicated. After the test drive, the technician should connect the scan tool to check for trouble codes and review the live data stream to verify that the injector pulse width and/or balance is even. Uneven injector balance can indicate incorrect installation procedures, cylinders that have low compression, or incorrectly uploaded injector codes.

COMPOSITE VEHICLE

1. You will notice as you read through the Task List that the job skills identified concentrate on the ability to diagnose, rather than repair. The panel of experts who developed the L2 test have identified three important general characteristics of drivability diagnosis.
2. Data is obtained from the vehicle using a variety of test instruments and is compared to known values obtained from the service manuals.
3. A good technician can draw valid conclusions from the relationship between published data and what he understands of the vehicles fuel, ignition, and emission-control systems.

The ASE panel of experts has developed a "Composite Vehicle" engine control system, which is described in detail in the "Composite Vehicle Preparation/Registration Booklet" you'll receive when you register for the test. The information is included here so that you can begin the familiarization process now.

The composite vehicle uses a "mass airflow" fuel-injection system design used by many domestic and import manufacturers. The system uses sensor, actuators, and control strategies that you should be familiar with from your shop experience. When you answer questions based on the Composite Vehicle, you will be simulating your real-world experience of using reference materials and test instruments to diagnose problems based on your knowledge of a particular engine-management system.

The following Composite Vehicle Information has been provided by the National Institute for Automotive Service Excellence (ASE). Delmar, Cengage Learning would like to thank ASE for providing this content for use in this study guide.

An electronic version of the Composite Vehicle reference booklet is also available at www.ase.com.

This booklet is intended only for reference when preparing for and taking the ASE Electronic Diesel Engine Diagnosis Specialist (L2) Test. The medium/heavy composite vehicle control system is based on designs common to many engine and vehicle manufacturers, but is not identical to any actual production engine or vehicle.

MEDIUM/HEAVY COMPOSITE VEHICLE INFORMATION

General Description

This generic in-line six cylinder diesel engine is equipped with a variable geometry turbocharger (VGT), charge air cooler (CAC), electronic unit injectors (EUI), closed crankcase ventilation, exhaust gas recirculation (EGR), and exhaust after treatment system. The engine rating is 400 hp at 1,800 rpm and develops a peak torque of 1450 lb. ft. at 1,200 rpm.

The Electronic Control Module (ECM)

The electronic control module is the microprocessor that receives electronic inputs from sensors and switches, and is calibrated to control fuel metering, injection timing, diagnostics, and engine protection. The ECM receives power from the battery and ignition switch and provides a reference voltage for some sensors. The software contained in the ECM determines how the electronic diesel engine control system operates. The ECM stores the calibration values that define rated horsepower, torque curves, and rpm specifications. The engine ECM communicates with other vehicle system control modules through the SAE J1939 data link (controller area network - CAN). The ECM is mounted on the engine.

Fuel System

The engine requires the use of ultra-low sulfur (ULSD) diesel fuel. Fuel supply (transfer) is provided by a positive displacement, mechanically driven pump. A preset spring-operated regulating valve mounted on the cylinder head fuel outlet gallery is used to maintain fuel pressure. Fuel delivery is provided by electronically controlled unit injectors (EUI). The injector solenoids are controlled by the ECM.

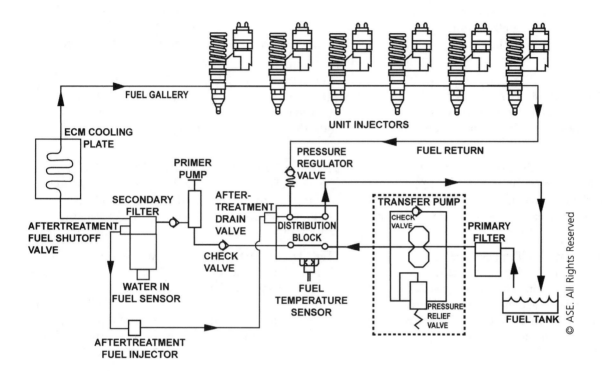

Engine Protection System

The engine protection system monitors coolant temperature, coolant level, oil temperature, intake manifold temperature, oil pressure, fuel temperature, EGR exhaust gas temperature, and diesel particulate filter (DPF) restriction. The system has three levels of protection: *warning, derate,* and *shutdown.*

In the *warning* mode, the yellow Check Engine Lamp (CEL) alerts the operator to a potential problem. In the *derate* (limp home) mode, the ECM gradually reduces engine power after illuminating the red Stop Engine Lamp (SEL). *Power derate* will begin 30 seconds after the initial *warning.* The *shutdown* mode will be indicated by a flashing SEL and activate for *engine coolant temperature, coolant level, engine oil pressure,* or *engine oil temperature* failures only. After the *warning* and *derate* modes have been activated, *shutdown* occurs when the condition reaches a preset value. The engine protection override switch can be used to temporarily delay the *shutdown* mode.

High Engine Coolant Temperature   The ECM will turn the engine cooling fan on at 205°F.

Engine Coolant Temperature	Power Derate/ Shutdown
225°F	20% Derate
230°F	40% Derate
235°F	60% Derate
240°F	Shutdown

High Fuel Temperature   A 20% power derate begins when the fuel temperature reaches 180°F. The derate increases to 40% if the temperature remains high.

High Intake Manifold Temperature   ECM will turn the engine cooling fan on at 190°F. A 20% power derate begins when the mainfold temperature reaches 210 °F. The derate increases to 40% if the temperature remains high.

High EGR Exhaust Gas Temperature   A 20% power derate begins when the EGR gas temperature reaches 525°F. The CEL is illuminated and the power derate increases to 40% when the temperature reaches 560°F. The SEL will not illuminate and the engine will return to full power when the temperature lowers.

Low Engine Coolant Level   After the initial warning, shutdown will occur if the coolant level remains low.

Low Engine Oil Pressure   Power derate/shutdown set points are dependent on the relationship between oil pressure and engine speed. If the oil pressure falls below these specifications, power derate will begin, and shutdown will occur if oil pressure continues to decrease.

	Engine Oil Pressure	
Engine Speed	Begin Derate	Shutdown
2000 rpm	40 psi	20 psi
1800 rpm	35 psi	17 psi
1400 rpm	25 psi	12 psi
1000 rpm	15 psi	8 psi
600 rpm	5 psi	2 psi

High Engine Oil Temperature   The ECM will turn the engine cooling fan on at 245°F. A 40% power derate begins when the oil temperature reaches 260°F. Shutdown occurs after 2 minutes if the temperature remains high.

High Diesel Particulate Filter Restriction   When DPF restriction reaches full soot load (level 3), the CEL is illuminated and a 40% derate begins. The SEL is illuminated and the power derate increases to 80% when the restriction reaches over-full soot load (level 4). A severely restricted DPF may cause the engine to shutdown.

Sensors

Accelerator Pedal Position Sensor (APP)   The APP contains two potentiometers that sense the position of the accelerator pedal. A reference voltage is sent to each potentiometer and as the angle of the accelerator pedal changes, the APP varies the signal voltages to the ECM. Both APP potentiometer voltages are compared by the ECM for diagnostics. If one or both of the signals are lost, the engine will not operate above idle speed.

Engine Oil Pressure Sensor (EOP)   The variable capacitance EOP sensor monitors engine oil pressure and is installed in the main lube oil gallery. The ECM uses this signal for engine protection and the instrument panel pressure gauge.

Intake Manifold Pressure Sensor (IMP)   The variable capacitance IMP sensor monitors intake manifold pressure and is located on the intake manifold. The ECM uses this signal to control fuel metering, injection timing, and turbocharger control.

Intake Manifold Temperature Sensor (IMT)   The thermistor IMT sensor monitors air temperature in the intake manifold. The ECM uses this signal to control fuel metering, injection timing, EGR operation, engine protection, and cooling fan operation.

Barometric Pressure Sensor (BARO)   The variable capacitance BARO sensor monitors ambient air pressure. The ECM uses this signal to adjust injection timing and fuel metering based on altitude.

Engine Coolant Temperature Sensor (ECT)   The thermistor ECT sensor monitors coolant temperature and is mounted in the engine block coolant jacket. The ECM uses this signal to control fuel management, engine protection, cooling fan operation, the instrument panel temperature gauge, and DPF regeneration.

Coolant Level Sensor (CLS)   The CL sensor monitors the level of the coolant in the radiator surge tank. The ECM uses this signal for engine protection when coolant is not detected.

Fuel Temperature Sensor (FTS)   The thermistor FT sensor monitors fuel temperature and is mounted on the fuel distribution block. The ECM uses this signal to adjust calculated fuel measurements to compensate for changes in fuel temperature and for engine protection.

Engine Oil Temperature Sensor (EOT)   The thermistor EOT sensor monitors engine oil temperature and is installed in the main lube oil gallery. The ECM uses this signal for engine protection.

Inlet Air Temperature Sensor (IAT)   The thermistor IAT sensor monitors air temperature in the inlet pipe before the turbocharger. The ECM uses this signal for fuel management.

Engine Position Sensor 1 (EPS1)   The Hall-effect EPS1 generates a digital signal that varies in frequency with the speed of the engine camshaft. The ECM uses the frequency and pulse width of this signal to determine camshaft position for fuel control and injection timing. The ECM will use this signal for calculated engine speed in the event of EPS2 (crankshaft) signal failure. The engine will shutdown in the event of EPS1 (camshaft) signal failure. The EPS1 is located in the engine front cover facing the trigger/tone wheel that is mounted on the camshaft gear.

Engine Position Sensor 2 (EPS2)   The Hall-effect EPS2 generates a digital signal that varies in frequency with the speed of the engine crankshaft. The ECM uses the frequency of this signal to determine engine speed for fuel control and injection timing. The ECM requires input from both EPS1 and EPS2 for starting. The EPS2 is located in the engine rear flywheel housing facing the trigger/tone wheel mounted on the crankshaft.

Exhaust Back Pressure Sensor (EBP)   The variable capacitance EBP sensor monitors exhaust pressure from a tube connected to the exhaust manifold. The ECM uses this signal for EGR valve and variable geometry turbocharger (VGT) operation.

Crankcase Pressure Sensor (CPS)   The variable capacitance CP sensor monitors pressure in the engine crankcase. The ECM uses this signal to verify the condition of the closed crankcase ventilation system and filter.

Water in Fuel Sensor (WIF)   The water in fuel sensor monitors the bottom of the primary fuel filter housing for the presence of water. When the water level covers the sensor probe, the reference voltage circuit is completed. The ECM uses this signal to alert the operator by illuminating the yellow WIF warning light on the instrument panel fuel gauge.

EGR Exhaust Temperature Sensor (EGRT)   The thermistor EGRT sensor monitors the temperature of the exhaust gases in the outlet of the EGR cooler. The ECM uses this signal for emissions management and engine protection.

EGR Differential Pressure Sensor (EGR Delta P)   The variable capacitance EGR Delta P sensor has two ports that monitor the exhaust gas pressure across the EGR differential pressure orifice. One port is located on each side of the EGR venturi. The ECM uses this pressure drop signal and the EGR temperature signal to calculate the amount of EGR flow into the intake manifold. The ECM commands the EGR valve and the VGT actuator positions to control the amount of exhaust gases entering the engine.

EGR Position Sensor (EGRP)   The three-wire fixed potentiometer EGRP sensor monitors the position of the EGR valve. The ECM uses this signal to verify proper operation of the EGR valve and motor for emission control and diagnostics.

Ambient Air Temperature Sensor (AT)   The thermistor AT sensor monitors the outside (ambient) air temperature and is mounted on the cab. The signal is sent from the body control module (BCM) to the ECM through the J1939 data link (CAN). The ECM uses this temperature for idle shutdown operation.

After treatment Fuel Pressure Sensor (AFP)   The variable capacitance AFP sensor monitors the pressure in the after treatment fuel system. The ECM uses this signal in determining operation of the after treatment fuel injector during regeneration. Active and stationary regeneration will be disabled if a fuel pressure fault is detected.

After treatment Diesel Oxidation Catalyst Inlet Temperature Sensor (EGT1)   The thermistor after treatment diesel oxidation catalyst (DOC) inlet temperature sensor monitors the exhaust gas temperature into the (DOC). The ECM uses this signal to control EGR and VGT actuator positions and after treatment fuel injection during DPF regeneration.

After treatment Diesel Particulate Filter Inlet Temperature Sensor (EGT2)   The thermistor after treatment diesel particulate filter (DPF) inlet temperature sensor monitors the exhaust gas temperature into the DPF. The ECM uses this signal in determining EGR and VGT actuator positions and after treatment fuel injection during DPF regeneration.

After treatment Diesel Particulate Filter Outlet Temperature Sensor (EGT3)   The thermistor after treatment diesel particulate filter (DPF) outlet temperature sensor monitors the exhaust gas temperature at the outlet of the DPF. The ECM uses this signal in determining EGR and VGT actuator position and after treatment fuel injection during DPF regeneration.

After treatment Diesel Particulate Filter Differential Pressure Sensor (DPF Delta P)
The variable capacitance DPF exhaust pressure sensor has two ports that monitor the exhaust gas pressure across the diesel particulate differential pressure orifice. One port is located on each side of the DPF. The ECM uses this pressure drop signal to calculate the amount of DPF restriction. Based on the amount of DPF restriction, the ECM will illuminate the series of warning lamps below:

- Level 1 – Low soot load. The DPF status lamp is illuminated.
- Level 2 – Moderate soot load. The DPF status lamp is flashing.
- Level 3 – Full soot load. The DPF status lamp is flashing and the yellow CEL is illuminated.
- Level 4 – Over-full soot load. The DPF status lamp is flashing and the red SEL is illuminated.

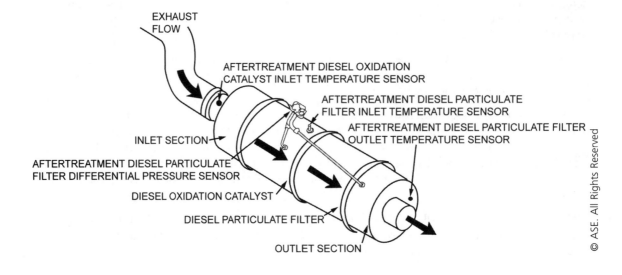

Switches

Engine Cooling Fan Manual Switch   This is a normally open switch that, when closed by the operator, requests that the ECM de-energize the engine cooling fan solenoid. This will cause the engine cooling fan to operate continuously.

Diagnostic ON/OFF Switch   This is a normally open momentary switch that, when pressed (closed) and released, requests the ECM to enter diagnostic mode. The yellow check engine lamp (CEL) will come ON and flash inactive fault codes. The red stop engine lamp (SEL) will come ON and flash active fault codes. If both lights momentarily turn ON and then OFF, no fault codes are present.

Air Conditioning High Pressure Switch   This normally closed pressure switch opens when the air conditioner high-side pressure reaches a preset value. This signals the ECM to operate the engine cooling fan.

Cruise Control ON/OFF Switch   This normally open switch enables cruise control operation when it is closed by the operator.

Cruise Set/Resume Switch   This switch has two momentary positions: SET/COAST and RESUME/ACCELERATE. Vehicle cruising speed can be set by pressing SET/COAST. Once a speed has been set, and then disengaged using the brake or clutch, the ECM will return to the previously set speed when RESUME/ACCELERATE is pressed. Vehicle cruise control set speed is adjusted lower by pressing SET/COAST and higher by pressing RESUME/ACCELERATE. The two set/resume switch positions are also used for engine speed settings during PTO operation.

Clutch Switch   This normally closed switch opens when the clutch pedal is depressed. This signals the ECM to cancel cruise control, power take-off mode, or engine braking. The switch input is also used to override the idle shutdown timer.

Service Brake Switch   This normally closed pressure switch opens when the service brake pedal is depressed. This signals the ECM to enable engine braking, cancel cruise control or cancel power take-off mode. The switch input is also used to override the idle shutdown timer.

Engine Brake ON/OFF Switch   This normally open switch signals the ECM that the operator is requesting engine brake system activation when closed. The ECM will energize the engine brake solenoids based on inputs from the accelerator pedal and the cruise control ON/OFF, clutch and service brake switches.

Engine Brake Selector Switch   This three-position switch sets the level of engine braking (LOW, MEDIUM, or HIGH). The engine brake solenoid for engine cylinders 3 & 4 is energized for the LOW level. The engine brake solenoids for cylinders 1 & 2 and cylinders 5 & 6 are energized for the MEDIUM level. All three solenoids are energized for the HIGH engine braking level.

Power Take-Off (PTO) ON/OFF Switch   This normally open switch enables PTO mode operation when closed. In this mode, the cruise control set/resume switch is used for two preset PTO engine speeds. SET/COAST is pressed for the first engine speed setting and RESUME/ACCELERATE is pressed for the second setting.

Power Take-Off (PTO) Remote Switch   This normally open switch is mounted outside the cab and enables remote PTO mode operation at a preset engine speed when closed.

Engine Protection Override Switch   This is a normally open momentary switch that, when pressed (closed) and released during the shutdown warning period, requests the ECM to override (delay) an engine protection shutdown for 30 seconds.

Diesel Particulate Filter Regeneration Switch This is a normally open momentary switch that, when pressed (closed) and released, requests the ECM to enable a stationary or parked DPF regeneration.

Parking Brake Switch This normally closed switch opens when the parking brake is released. The switch position signal is sent from the body control module (BCM) to the ECM through the J1939 data link (CAN). The ECM uses this input to control stationary or parked DPF regeneration, idle shutdown, and PTO operation.

Actuators

Electronic Unit Injectors (EUI) These electromechanical diesel fuel injectors contain a solenoid controlled by the ECM to manage fuel timing and metering. The ECM injector driver circuitry supplies high voltage to the injector solenoids which are energized by controlling the ground circuits within the ECM. Injector calibration codes need to be programmed into the ECM to compensate for manufacturing tolerances.

Engine Cooling Fan Solenoid When energized by the ECM, this normally closed solenoid supplies air pressure to disengage the engine cooling fan clutch and turn the fan OFF. The solenoid can be de-energized by the ECM, shutting off air pressure to engage the fan clutch and turn the fan ON. The ECM engages the fan to assist with engine braking when the selector switch is in the HIGH position. The ECM will operate the fan for engine protection when coolant temperature reaches 205°F, engine oil temperature at 245°F, intake manifold temperature at 190°F, and when the A/C high-pressure switch opens.

Stop Engine Lamp The red stop engine lamp (SEL) illuminates when the engine protection system is in the derate and shutdown modes. The red SEL can also be used to read active fault "flash" codes when in the diagnostic mode.

Check Engine Lamp The yellow check engine lamp (CEL) illuminates when the engine protection system is in the warning mode, or when electronic control system failures are occurring and an active fault is present. The yellow CEL can also be used to read inactive fault "flash" codes when in the diagnostic mode. The CEL and SEL are used in combination with the after treatment regeneration status lamp (DPFR) for DPF restriction status and regeneration requirements.

Engine Brake Solenoids These three solenoids can be energized by the ECM to provide engine braking on two, four or all six cylinders. The engine brake can be activated during cruise control operation. The solenoids will be energized after the set speed has been exceeded or when the service brake pedal is depressed and the accelerator pedal is released. The ECM will not energize the solenoids when the clutch pedal is depressed.

EGR Valve Motor The EGR valve stepper motor is mounted on the EGR valve and controlled by the ECM through the J1939 data link (CAN). Exhaust gases flow through the EGR cooler to the EGR valve and into a venturi (mixing chamber combining intake air and exhaust gas) at the intake manifold. The ECM will disable the EGR valve (default closed position) in the event of motor data link communication failure.

Variable Geometry Turbocharger Actuator The VGT actuator is mounted on the turbocharger and controlled by the ECM through the J1939 data link (CAN). The actuator operates an internal sliding nozzle ring in the turbine housing. The sliding nozzle ring allows for control of turbine shaft speed by regulating exhaust gas flow through the turbocharger. As the ring is moved from the open to the closed position, exhaust gas flow velocity increases. The actuator has self-diagnostic capabilities and has a preset default position in the event of actuator failure. The ECM calculates turbocharger shaft speed for displayed data and diagnostics.

(Actuator section continues after the following wiring diagrams.)

DIESEL ENGINE ELECTRONIC CONTROLS-1

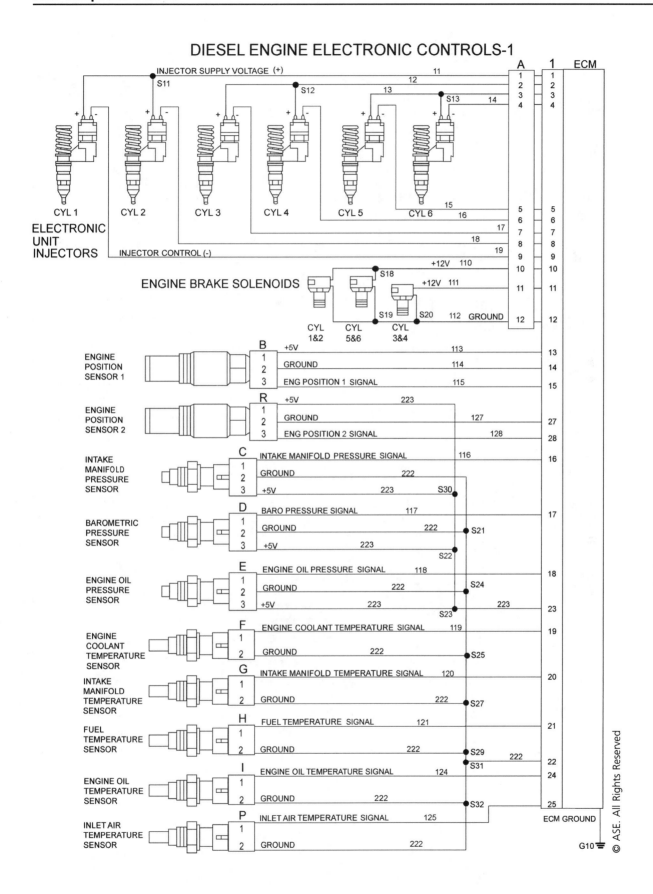

DIESEL ENGINE ELECTRONIC CONTROLS-2

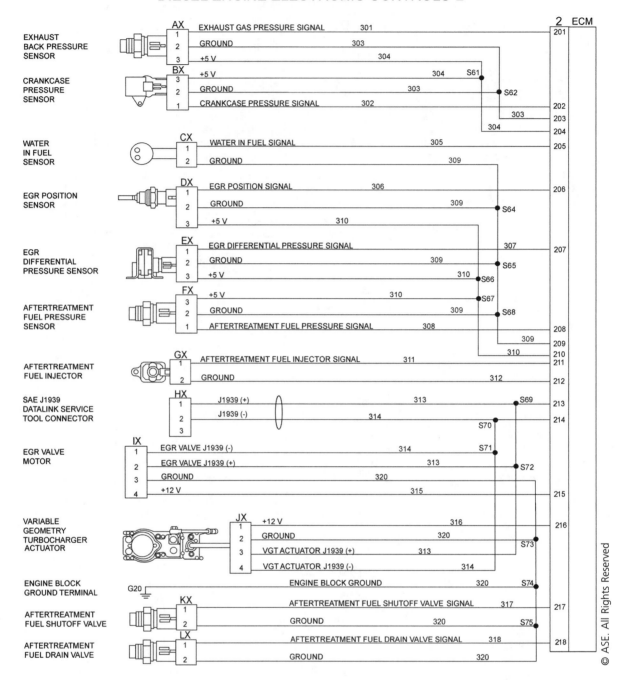

VEHICLE OEM ELECTRONIC CONTROLS-3

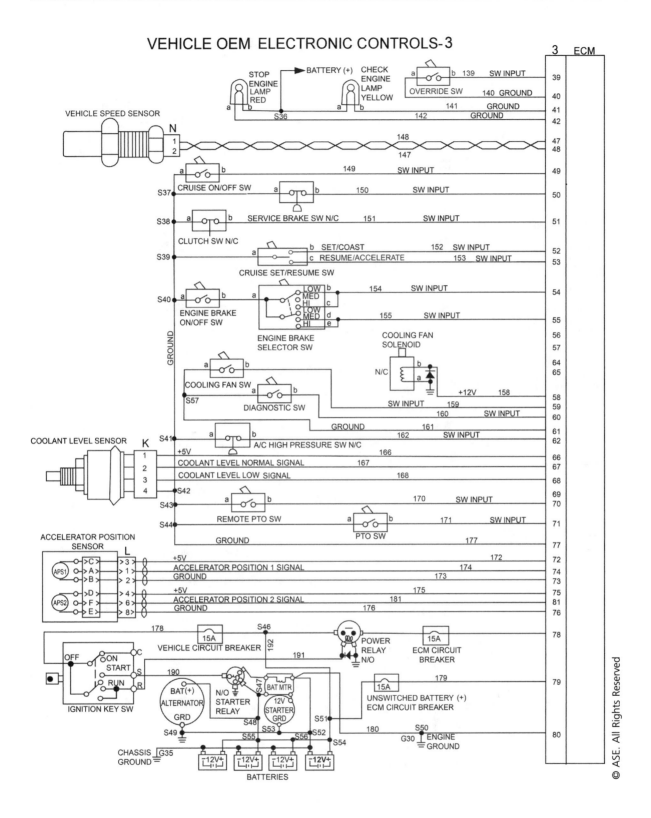

VEHICLE OEM ELECTRONIC CONTROLS - 4

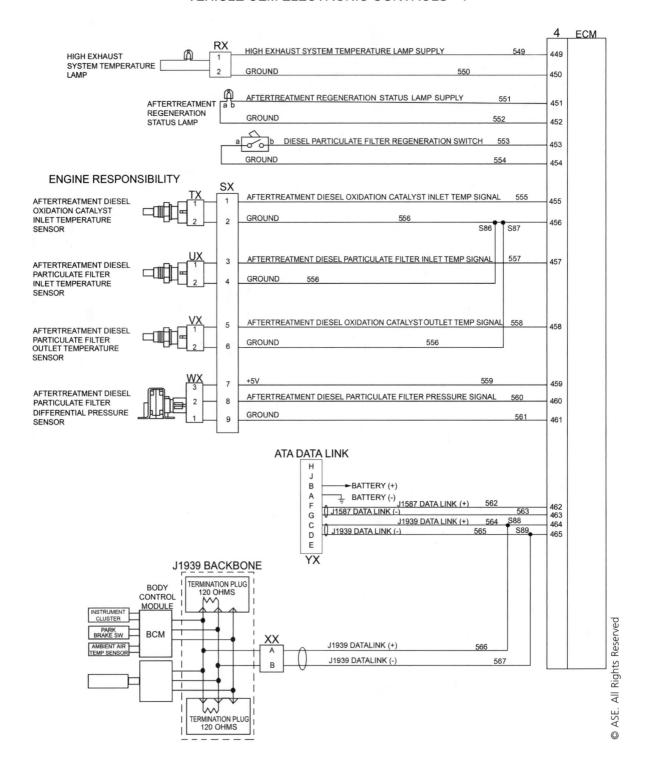

Idle Shutdown Timer (IST)   The idle shutdown feature reduces the amount of fuel burned and increases engine life by shutting down the engine after a period of engine idling with no driver activity. Thirty (30) seconds before the shutdown occurs, the stop engine lamp flashes to alert the driver of an impending shutdown. The driver can override the shutdown by depressing the service brake, clutch, or accelerator pedal during the warning period. If the override is successful, the SEL will continue flashing for two minutes. The idle shutdown time period will restart when the idle condition is detected by the ECM.

Idle shutdown can interact with the PTO feature. It can cause the engine to shutdown when in PTO mode. If the idle shutdown percent load threshol is **not** exceeded, the engine will be shutdown.

The following conditions must be set for the idle shutdown timer to activate. Any change to one or more of these conditions will reset or disable the IST.

Enable Conditions for IST:

- Engine is idling below 750 rpm.
- Vehicle speed is 0 mph.
- No active vehicle speed sensor diagnostic faults.
- PTO/Remote PTO is operating below the percent load threshold. (Customer Programmable Parameters)
- Ambient air temperature is between 40°F and 80°F (CAN message from body control module). (Customer Programmable Parameters)
- No active inlet air temperature sensor diagnostic faults.
- Engine coolant temperature is above 140°F.
- No active engine coolant temperature sensor diagnostic faults.
- Stationary diesel particulate filter (DPF) regeneration is inactive.
- Accelerator pedal position released (at idle).
- Service brake switch is closed.
- Clutch switch is closed.
- Parking brake applied (CAN message from body control module).

Exhaust After treatment

High Exhaust System Temperature Lamp (HEST)   The (HEST) lamp is illuminated by the ECM when the exhaust gas temperatures monitored by EGT3 exceed 850°F and vehicle speed is below 5 mph.

After treatment Regeneration Status Lamp (DPFR)   The ECM illuminates the after treatment regeneration status lamp when DPF restriction reaches set points based on the input from the DPF differential pressure sensor (DPF Delta P). This indicates the need for DPF regeneration based on the levels shown below:

- Level 1 – Low soot load. The DPFR status lamp is illuminated.
- Level 2 – Moderate soot load. The DPFR status lamp is flashing.
- Level 3 – Full soot load. The DPFR status lamp is flashing and CEL is illuminated.
- Level 4 – Over-full soot load. The DPFR status lamp is flashing and the SEL is illuminated.

After treatment Fuel Injector (AFI)   Fuel transfer pressure is supplied to the pulse width modulated (PWM) after treatment fuel injector from a shutoff valve located on the secondary fuel filter outlet. The ECM injects diesel fuel into the exhaust gas, upstream of the diesel oxidation catalyst (DOC), to raise the temperature. When DOC inlet temperature is 600°F, the ECM begins operating the AFI and the duty cycle is varied until the exhaust temperature increases to the desired level for regeneration.

After treatment Fuel Shutoff Valve (AFS)   The AFS controls the fuel flow to the supply valve on the secondary fuel filter housing when commanded by the ECM. After startup and during the first minute of engine operation, the ECM performs a self-test on the system by opening the AFS to pressurize the after treatment fuel system. The AFD is then opened to relieve the fuel pressure while the ECM monitors the AFP sensor readings. If a problem is detected, the ECM will disable active and stationary regeneration until the next key cycle.

After treatment Fuel Drain Valve (AFD)   The AFD valve is used to maintain and relieve the fuel pressure in the after treatment fuel system. The ECM commands the AFD to open and allow fuel to flow into the fuel return line.

After treatment Regeneration System Operation   The after treatment diesel particulate filter accumulates soot and ash during engine operation. Soot is oxidized and removed during regeneration. Ash accumulates in the DPF over the service life of the unit. The DPF needs to be disassembled and the ash is removed by a special cleaning process.

Regeneration is passive or active based on engine operating conditions, DPF restriction level, and driver's response requirement. The ECM will not enable active or stationary regeneration if the DPF restriction is at Level 4.

Passive regeneration occurs when the exhaust temperatures are high enough through normal engine operation. This typically happens when the vehicle is driven at highway speeds and/or under heavy loads.

Active regeneration occurs when the exhaust temperatures are *not* high enough to oxidize the soot collected in the DPF. This will occur more frequently in vehicles with low speed and low load duty cycles. The ECM will inject diesel fuel into the exhaust gas before the inlet of the diesel oxidation catalyst (DOC) to raise the temperature for regeneration. The ECM will enable and disable active regeneration as needed. The speed threshold for active regeneration to take place is 25 mph and it will stop when the vehicle speed drops below it. The exhaust temperature during active regeneration can reach up to 1,500°F.

Stationary (Parked) regeneration is needed when conditions are *not* reached during vehicle operation. This is a form of <u>active</u> regeneration initiated by the operator using the DPF Regeneration Switch when the vehicle is not moving. The vehicle must be parked with the transmission in NEUTRAL and the parking brake set. There can be no input from the accelerator, brake, and clutch pedal. The ECM controls the regeneration process that can last up to 1½ hours depending on the amount of DPF restriction.

Data Link Communications

The SAE J1939 data link bus (controller area network - CAN) allows the ECM to communicate with other vehicle control systems such as transmission, automatic traction control, antilock brake, and body controllers. The J1939 data link is an unshielded twisted pair (UTP). The ECM will broadcast public data link messages when the key switch is in the ON position and stops when the key switch is OFF. The ECM will also broadcast private data link messages to drive actuators using J1939 protocol during engine operation and diagnostic programming. The Society of Automotive Engineers (SAE) recommends a maximum backbone harness of 131 feet (40 meters) in length. The harness is terminated at each end with a 120 ohm resistor. Up to 30 different devices can be attached to the J1939 backbone harness at one time. Each device is connected to the backbone harness with a 3 pin stub connector and can be a maximum of 3.3 feet (1 meter) in length.

Any of the following conditions will cause the data communications bus to fail and result in the storage of network DTCs: either data line is shorted to power, to ground, or to the other data line.

The data bus will remain operational when one of the two modules containing a terminating resistor is not connected to the network. The data bus will fail when both terminating resistors are not connected to the network. Data communication failures do not prevent the ECM from providing fuel management.

The Diagnostic Tool communicates with the ECM through the 9 pin ATA connector using J1708 protocols over the J1939 and/or J1587 data links. This allows for diagnostic information retrieval and parameter calibration setting. The 9 pin ATA data link connector is located in the cab.

Diagnostic Trouble Codes (DTC)

Trouble codes can be active or inactive. Active codes indicate that the problem currently exists and inactive codes indicate that a problem once existed. Flash codes represent digits assigned to diagnostic trouble codes, so that DTCs can be retrieved through the diagnostic lamps. Active codes are indicated by the red stop engine lamp (SEL). Inactive codes are indicated by the yellow check engine lamp (CEL). When using the diagnostic tool, DTCs are formatted under SAE J1939 standards and descriptions.

Diagnostic Equipment

A *breakout tool* can be connected into a circuit to measure voltage signals and resistance values with a *digital multimeter* (DMM). A *diagnostic tool* can be connected to the data link connector to read engine data, diagnostic codes, and set programmable parameters. ECM software updates (re-flash) can be performed using a PC based diagnostic tool.

The Displayed Data chart shows how *diagnostic tool* data will be presented in some Composite Vehicle test questions. The chart includes the normal operating range for senor voltages and how the status of components (switches and lamps) or operational modes will be indicated. The minimum – maximum measurement range and values for engine data (temperatures, pressures, and speeds) is also shown.

(Please refer to the chart that follows.)

Displayed Data

Displayed Data	Value Range	Displayed Data	Value Range
After treatment Diesel Oxidation Catalyst Inlet Temperature	0 – 2000°F	Engine Coolant Level	Low/Normal
After treatment Diesel Oxidation Catalyst Inlet Temperature Sensor Signal Voltage	.5 – 4.5 V	Engine Coolant Temperature	–40° – 248°F
After treatment Diesel Particulate Filter Differential Pressure	0 – 10 inHg	Engine Coolant Temperature Sensor Signal Voltage	.5 – 4.5 V
After treatment Diesel Particulate Filter Differential Pressure Sensor Signal Voltage	.5 – 4.5 V	Engine Oil Pressure	0 – 125 psi
After treatment Diesel Particulate Filter Inlet Temperature	0 – 2000°F	Engine Oil Pressure Sensor Signal Voltage	.5 – 4.5 V
After treatment Diesel Particulate Filter Inlet Temperature Sensor Signal Voltage	.5 – 4.5 V	Engine Oil Temperature	–40° – 300°F
After treatment Diesel Particulate Filter Lamp Status	On/Off	Engine Oil Temperature Sensor Signal Voltage	.5 – 4.5 V
After treatment Diesel Particulate Filter Outlet Temperature	0 – 2000°F	Engine Protection Shutdown	On/Off
After treatment Diesel Particulate Filter Outlet Temperature Sensor Signal Voltage	.5 – 4.5 V	Engine Speed	0 – 3000 rpm
After treatment Diesel Particulate Filter Regeneration Start Switch Status	On/Off	Exhaust Gas Pressure (EBP)	0 – 125 InHg
After treatment Fuel Pressure	0 – 300 psi	Exhaust Gas Pressure Sensor Signal Voltage	.5 – 4.5 V
After treatment Fuel Pressure Sensor Signal Voltage	5 – 4.5 V	Fan Control Switch	On/Off
After treatment High Exhaust System Temperature Lamp Status	On/Off	Fuel Temperature Sensor Signal Voltage	.5 – 4.5 V
Air Conditioning Pressure Switch	On/Off	Inlet Air Temperature	–40° – 248°F
Amber/Yellow Warning Light Status (CEL)	On/Off	Inlet Air Temperature Sensor Signal Voltage	.5 – 4.5 V
Barometric Air Pressure	0 – 30 inHg	Intake Manifold Air Temperature	–4° – 248°F
Barometric Air Pressure Sensor Signal Voltage	.5 – 4.5 V	Intake Manifold Air Temperature Sensor Signal Voltage	.5 – 4.5 V
Battery Voltage	0 – 15 V	Intake Manifold Pressure (Gauge)	0 – 60 inHg
Brake Pedal Position Switch	Released/Depressed	Intake Manifold Pressure Sensor Signal Voltage	.5 – 4.5 V
Clutch Pedal Position Switch	Released/Depressed	Key Switch	On/Off
CMP/Engine Position 1 Signal	Yes/No	Percent Accelerator Pedal (APP)	0 – 100%
CKP/Engine Position 2 Signal	Yes/No	PTO Decrement Switch	On/Off
Crankcase Pressure	0 – 40 inH20	PTO Increment Switch	On/Off
Crankcase Pressure Sensor Signal Voltage	.5 – 4.5 V	PTO On/Off Switch	On/Off
Cruise Control On/Off Switch	On/Off	PTO Status	Active/Inactive
Cruise Control Set/Resume Switch	Set/Neutral/Resume	Red Stop Lamp Status (SEL)	On/Off
Diagnostic Switch	On/Off	Remote PTO Switch	On/Off
ECM Time (Key On Time)	HH:MM:SS	Sensor Supply 1	4.75 – 5.25 V
EGR Differential Pressure	0 – 10 inHg	Sensor Supply 2	4.75 – 5.25 V
EGR Differential Pressure Sensor Signal Voltage	.5 – 4.5 V	Sensor Supply 3	4.75 – 5.25 V
EGR Temperature	–40° – 400°F	Synchronization State	Yes/No
EGR Temperature Sensor Signal Voltage	.5 – 4.5 V	Transmission Gear Ratio	.1 – 16
EGR Valve Position Commanded	0 – 100%	Turbocharger Actuator Position Measured (Percent Closed)	0 – 100%
EGR Valve Position Sensor Signal Voltage	.5 – 4.5 V	Turbocharger Actuator Position Sensor Signal Voltage	.5 – 4.5 V
Engine Brake On/Off Switch	On/Off	Turbocharger Speed	0 – 200,000 rpm
Engine Brake Selector Switch	Low/Med/High	Vehicle Speed	0-127 mph

Programmable Parameters

Programmable parameters are the specifications that can be set within the ECM to control operating functions. The parameters are stored in non-volatile memory. A customer password is available for programming protection. A list of parameter ranges and their settings are shown below.

Feature	Range	Setting	Feature	Range	Setting
Road Speed Governor			**PTO/Remote PTO**		
Accelerator Max. Road Speed	30–120 mph	65 mph	Max PTO Speed	600–2500 rpm	1000 rpm
Accelerator Upper Droop	0–3 mph	0 mph	Min PTO Speed	600–2500 rpm	700 rpm
Accelerator Lower Droop	0–3 mph	1 mph	Set PTO Speed	600–2500 rpm	900 rpm
Global Max. Road Speed	0–120 mph	120 mph	Resume PTO Speed	600–2500 rpm	1000 rpm
Gear Down Protection (GDP)			Remote PTO Speed	600–2500 rpm	1000 rpm
GDP Light Load Vehicle Speed	30–1000 mph	54 mph	Max. Engine Load	0–1850 ft.lb.	800 ft.lb.
GDP Heavy Load Vehicle Speed	30–1000 mph	60 mph	Max. Vehicle Speed	0–30 mph	0 mph
Idle Speed Control			Ramp Rate	100–2500 rpm/sec	250 rpm/sec
Idle Engine Speed	600–850 rpm	700 rpm	**After treatment**		
Idle Shutdown			Stationary Regeneration in PTO	Enabled/Disabled	Enabled
Idle Shutdown Timer	1–100 min.	5.0 min.	Automatic Stationary Regeneration	Enabled/Disabled	Enabled
Idle Shutdown Lower Ambient Air Temp.	0–100°F	40°F	Mobile/Active Regeneration	Enabled/Disabled	Enabled
Idle Shutdown Upper Ambient Air Temperature Override	0–100°F	80°F	Minimum Vehicle Speed	20–100 mph	25 mph
Idle Shutdown Percent PTO Load Override	0–100%	100%	Diesel Particulate Filter Lamp	Enabled/Disabled	Enabled
Idle Shutdown	Enabled/Disabled	Enabled	Diesel Particulate Filter Regeneration Permit Switch	Enabled/Disabled	Enabled
Idle Shutdown Manual Override	Enabled/Disabled	Enabled	Diesel Particulate Filter Regeneration Start Switch	Enabled/Disabled	Enabled
Fan Control			High Exhaust System Temperature Lamp	Enabled/Disabled	Enabled
Minimum Fan On Time	0–1000 sec.	240 sec.	**Vehicle Setup Parameters**		
Fan Control Clutch Logic		0 Volts ON	Rear Axle Ratio	2-15.98	4.1
Fan On During Engine Braking	Enabled/Disabled	Enabled	Tire Size	301–700 rev/mile	501
Fan Vehicle Speed Interaction	Enabled/Disabled	Enabled	No. of Tailshaft Gear Teeth	1–64	16
Fan Control A/C Press Switch	Enabled/Disabled	Enabled	Vehicle Speed Sensor Type		Magnetic
Fan Control	Enabled/Disabled	Enabled	Max. Engine Speed w/out VSS	1400–3000 rpm	1800 rpm
Cruise Control/Engine Brakes			Max. Engine Speed with VSS	1400–3000 rpm	2100 rpm
Max. Cruise Control Speed	30–102 mph	65 mph	Trans. Top Gear Ratio	0.1-2	1.00
Cruise Control Upper Droop	0–3 mph	0 mph	Trans. One Gear Down Ratio	0.1-16	1.34
Cruise Control Lower Droop	0–3 mph	2 mph	Transmission Type		Manual
Cruise Control Speed Delta for Max. Engine Brake	0–6 mph	5 mph	Coolant Level	Enabled/Disabled	Enabled
Cruise Control Speed Delta for Min. Engine Brake	0–102 mph	3 mph	Water in Fuel Sensor	Enabled/Disabled	Enabled
Cruise Control Feature	Enabled/Disabled	Enabled	Multiplexing Parking Brake Switch		J1939
Engine Brake Cruise Control Activation	Enabled/Disabled	Enabled	Parking Brake Source Address	0–255	49
Engine Brake Min. Vehicle Speed	0–35 mph	0 mph	**Engine Protection**		
Engine Brake Delay	0–10 sec.	0.0 sec.	Engine Protection Shutdown Feature	Warning/Derate/ Shutdown	Shutdown
Engine Brake Service Brake Activation	Enabled/Disabled	Enabled	Engine Protect Restart Inhibit	Enabled/Disabled	Enabled
Engine Brake Control	Enabled/Disabled	Enabled	Manual Override	Enabled/Disabled	Enabled

Engine Operating Specifications

Fuel Supply Pressure

Cranking: 20 psi. min.

Rated rpm: 90–100 psi.

Intake Manifold Pressure

Rated rpm: 30–45 psi.

Engine Oil Pressure

Idle: 15–35 psi. min.

Rated rpm: 50–70 psi.

Electrical Specifications

J1587 Data Link

Positive Wire to chassis ground: 2.5 to 5.0 VDC

Negative Wire to chassis ground: 0.0 to 2.5 VDC

J1939 Data Link

Positive wire to negative wire
50 to 70 ohms

Terminal Resistance
110 to 130 ohms

ECM

9.0 to 16.0 VDC supply voltage

Maximum voltage drop on all circuits (except injectors): 0.5 VDC

ECM 5V-Reference Sensor Supply Groups

Circuit Group	ECM Pin Numbers
Sensor Supply 1	13, 23, 66, 204
Sensor Supply 2	459, 210
Sensor Supply 3	72, 75

5 V Reference Power Supply

At ECM: 4.75 to 5.25 VDC

At Harness: 4.75 to 5.25 VDC

All Shorts to External Voltage

OK: if less than 1.5 VDC

All Shorts to Ground

OK: if greater than 100K ohms

(No short circuit)

All Continuity Checks

OK: if less than 10 ohms

(No open circuit)

Fuel Injectors

0.5 to 5.0 ohms resistance

100 to 120 VDC supply voltage

Sensor Specifications

Vehicle Speed Sensor

Coil Resistance: 1100 to 1500 ohms

Intake Manifold Pressure Sensor

Pressure psig	inHg	Voltage VDC
atmospheric	atmospheric	0.50
12.5	25.45	1.50
25.0	50.90	2.50
37.5	76.35	3.50
50.0	101.80	4.50

All Temperature Sensors

Resistance: 600 to 1600 ohms

Barometric Pressure Sensor

Pressure inHg	Altitude feet	Voltage VDC
29.9	0 (sea level)	4.00 – 4.50
26.8	3000	3.50 – 4.30
24.0	6000	3.10 – 3.90
21.4	9000	2.90 – 3.70
19.1	12000	2.50 – 3.30

Engine Oil Pressure Sensor

Pressure		Voltage VDC
kPa	psi	
0	0	0.50
172	25	1.50
345	50	2.50
517	75	3.50
689	100	4.50

All Temperature Sensors (except exhaust after treatment and EGR)

Temperature		Voltage VDC
°C	°F	
120	248	0.50
82	180	0.75
54	130	1.50
32	90	2.50
0	32	3.60
−40	−40	4.50

After treatment Fuel Pressure Sensor

Pressure		Voltage VDC
kPa	psi	
0	0	0.50
345	50	1.50
689	100	2.50
1034	150	3.50
1379	200	4.50

Coolant Level Sensor

Coolant present: CL Normal signal = 5 VDC, CL
Low signal near 0 VDC

Coolant NOT present: CL Low signal = 5 VDC, CL
Normal signal near 0 VDC

Accelerator Pedal Position Sensor

Accelerator Pedal Position % depressed	APP 2 Sensor Voltage	APP 1 Sensor Voltage
0	0.50	1.50
5	0.65	1.65
10	0.80	1.80
15	0.95	1.95
20	1.10	2.10
25	1.25	2.25
40	1.70	2.70
50	2.00	3.00
60	2.30	3.30
75	2.75	3.75
80	2.90	3.90
100	3.50	4.50

EGR Position Sensor

EGR Valve (% Open)	Sensor Voltage
0	0.50
25	1.50
50	2.50
75	3.50
100	4.50

All Sensor Signal Voltages

Out of Range Low = Less than 0.20 VDC

Normal Range = 0.5 – 4.5 VDC

Out of Range High = Greater than 4.80 VDC

INTRODUCTION

Included in this section are a series of six individual preparation exams that you can use to help determine your overall readiness to successfully pass the Electronic Diesel Engine Diagnosis Specialist (L2) ASE certification exam. Located in Section 7 of this book you will find blank answer sheet forms you can use to designate your answers to each of the preparation exams. Using these blank forms will allow you to attempt each of the six individual exams multiple times without risk of viewing your prior responses.

Upon completion of each preparation exam, you can determine your exam score using the answer keys and explanations located in Section 6 of this book. Included in the explanation for each question is the specific task area being assessed by that individual question. This additional reference information may prove useful if you need to refer back to the task list located in Section 4 for additional support.

PREPARATION EXAM 1

1. A vehicle has been brought into the shop for diagnosis. Technician A says the engine serial number can be found on a data tag mounted on the engine. Technician B says the engine serial number is the last six digits of the vehicle identification number (VIN). Who is correct?

   A. A only
   B. B only
   C. Both A and B
   D. Neither A nor B

2. Technician A says the antilock brakes/electronic stability control systems (ABS/ESC) can cause the engine to derate. Technician B says the ABS/ESC system can cause the engine to shut off. Who is correct?

   A. A only
   B. B only
   C. Both A and B
   D. Neither A nor B

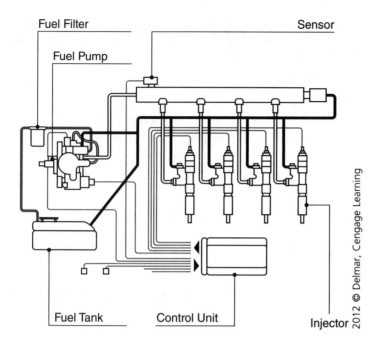

3. Referring to the figure above, which fuel system is illustrated?

    A. Electronic unit injector (EUI)

    B. Hydraulically actuated electronic unit injectors (HEUIs)

    C. Pump line nozzle electronic (PLN-E)

    D. Common rail

4. A connector lock is broken. Which of the following is the LEAST LIKELY acceptable repair method?

    A. Replace only the lock.

    B. Replace the lock and connector shell.

    C. Use a nylon tie strap to secure the two halves of the connector together.

    D. Replace the entire connector.

5. Which of the following would be the LEAST LIKELY cause of high air intake temperature?

    A. A restricted charge air cooler

    B. A restricted radiator

    C. A restricted condenser

    D. A restricted air filter

6. Technician A says that the customer complaint should be verified before performing any work. Technician B says that verifying the complaint may include talking to the customer or driving the vehicle.

    A. A only

    B. B only

    C. Both A and B

    D. Neither A nor B

7. A high-mileage engine has low power and a misfire on cylinder #5. The technician finds an open injector coil on cylinder #5. After installing all new injectors, the engine has a misfire on cylinder #3. Which of the following could be the cause?

    A. Low compression on cylinder #5
    B. Low compression on cylinder #3
    C. A damaged injector electrical pass-through connector on cylinder #3
    D. A damaged injector electrical pass-through connector on cylinder #5

8. All of the following are styles of APP sensors EXCEPT:

    A. A single potentiometer.
    B. A triple potentiometer.
    C. Optical.
    D. Hall effect.

9. Which of the following would be the last step in a complete diagnosis and repair of a diesel engine?

    A. Replace the faulty component.
    B. Research service literature.
    C. Verify the repair.
    D. Verify the complaint.

10. Which of the following is LEAST LIKELY to cause a no-start condition?

    A. Incorrect idle speed programmed into the ECM
    B. A failed engine position sensor
    C. A failed ECM
    D. Incorrect engine calibration files in the ECM

11. When reprogramming an ECM, all of the following are important safety precautions EXCEPT:

    A. The scan tool/laptop should be connected to a 110 volt AC (VAC) current.
    B. Primary air pressure on the truck should be built to 120 psi.
    C. The truck batteries should be fully charged.
    D. The technician should verify that the latest flash file is being used.

12. Each of the following would be considered a mechanical engine problem EXCEPT:

    A. Low compression.
    B. Clogged diesel particulate filter (DPF).
    C. Leaking exhaust gas recirculation (EGR) cooler.
    D. Open injector wiring harness.

13. Technician A says a failure of the data bus could prevent the automated manual transmission from shifting properly. Technician B says a failure of the data bus could prevent the scan tool from communicating with the transmission controller. Who is correct?

    A.　A only
    B.　B only
    C.　Both A and B
    D.　Neither A nor B

14. Technician A says air inlet restriction should only be measured at idle. Technician B says a water manometer is used to measure air inlet restriction. Who is correct?

    A.　A only
    B.　B only
    C.　Both A and B
    D.　Neither A nor B

15. Technician A says a damaged throttle position sensor (TPS) can prevent proper EGR operation. Technician B says a damaged TPS can cause a no-start condition. Who is correct?

    A.　A only
    B.　B only
    C.　Both A and B
    D.　Neither A nor B

16. An engine equipped with an HEUI fuel system fails to start. A technician connects a scan tool to the engine. While cranking, the injection actuation pressure is 270 psi. Technician A says the fuel transfer pump is not providing sufficient pressure. Technician B says that the fuel filters should be checked for restriction. Who is correct?

    A.　A only
    B.　B only
    C.　Both A and B
    D.　Neither A nor B

17. A diesel engine has several unrelated DTCs. During diagnosis, the technician measures AC voltage at the back of the alternator. Which of the following would be considered an acceptable voltage?

    A.　13.2 VAC
    B.　12.9 VAC
    C.　1.3 VAC
    D.　0.3 VAC

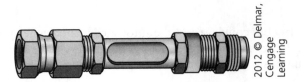

2012 © Delmar, Cengage Learning

18. Referring to the figure above, the tool shown is used to:

    A. Measure diesel fuel-specific gravity.

    B. Measure coolant-specific gravity.

    C. Test the fuel system for air.

    D. Test the air intake system for diesel fuel.

19. A vehicle has been brought to the shop multiple times to have the DPF cleaned using a stationary regeneration. Which of the following is the LEAST LIKELY cause?

    A. Incorrect ECM programming for the active regeneration of the after treatment device

    B. The use of ultralow-sulfur fuel

    C. The use of high-sulfur fuel

    D. Incorrect programming of the minimum vehicle speed parameter for active regeneration

20. Technician A says EUI injectors will need to be adjusted after installation. Technician B says that when replacing EUI injectors, the injector calibration code may need to be entered into the engine ECM. Who is correct?

    A. A only

    B. B only

    C. Both A and B

    D. Neither A nor B

21. Where would the after treatment diesel fuel injector be located?

    A. In the exhaust manifold before the turbocharger

    B. In the exhaust system after the turbocharger

    C. In the exhaust system between the diesel oxidation catalyst (DOC) and the DPF

    D. In the intake system prior to the EGR valve

22. Technician A says glow plugs can be checked with an ohmmeter. Technician B says air inlet heaters can be checked with an amp clamp. Who is correct?

    A. A only

    B. B only

    C. Both A and B

    D. Neither A nor B

23. Each of the following is true concerning the exhaust after treatment system EXCEPT:

    A. It converts soot to ash.

    B. It may need to be removed from the truck to clean soot from the DPF.

    C. It needs to be removed from the truck to clean ash from the DPF.

    D. The ECM monitors the DPF for restriction with a differential pressure sensor.

24. All of the following are methods of DPF regeneration EXCEPT:

    A. Passive.

    B. Active.

    C. Stationary.

    D. Engine off.

25. A diesel engine has had the second set of injectors replaced in less than 10,000 miles. Both sets of injectors had excessive internal leakage. Technician A says the problem could be a restricted air filter. Technician B says the problem could be a restricted DPF. Who is correct?

    A. A only

    B. B only

    C. Both A and B

    D. Neither A nor B

26. Refer to the composite vehicle to answer this question: The composite diesel engine blows black smoke under acceleration. Which of the following is the LEAST LIKELY cause?

    A. A damaged DPF

    B. A restricted exhaust

    C. Worn injectors

    D. Incorrect injector calibration codes

27. Refer to the composite vehicle to answer this question: The technician has retrieved a diagnostic trouble code (DTC) for a failed DPF. When checking the voltage with the engine running under various loads and RPMs, the after treatment DPF differential pressure sensor signal voltage never changes from 0.8 volts. Which of the following could be the cause?

    A. A restricted DPF

    B. A missing DPF

    C. A restricted EGR passage

    D. A stuck EGR valve

28. Refer to the composite vehicle (diagram on page 40 of the booklet) to answer this question: The ECM will not communicate with the transmission module. Which data bus could be the cause?

    A. J1708

    B. J1587

    C. J1922

    D. J1939

29. Refer to the composite vehicle to answer this question: Technician A says the idle shutdown timer is an adjustable parameter on the composite engine. Technician B says that maximum engine RPM is an adjustable parameter on the composite engine. Who is correct?

    A. A only

    B. B only

    C. Both A and B

    D. Neither A nor B

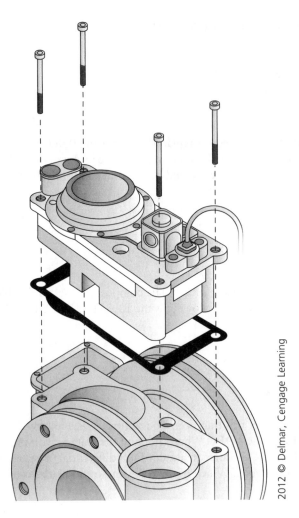

2012 © Delmar, Cengage Learning

30. Refer to the composite vehicle for this question: Referring to the figure above, what component is being removed?

   A.   The after treatment injector

   B.   The variable geometry turbocharger (VGT) actuator

   C.   The boost pressure sensor

   D.   The DPF

31. Refer to the composite vehicle to answer this question: The EGR valve fails to operate. There is a signal voltage to the valve at the IX connector. Technician A says the EGR cooler may be restricted. Technician B says the EGR venturi may be restricted. Who is correct?

   A.   A only

   B.   B only

   C.   Both A and B

   D.   Neither A nor B

32. Refer to the composite vehicle to answer this question: A vehicle has been sitting in the shop for more than 24 hours. A technician connects a scan tool to the engine and finds the coolant temperature and air inlet temperature to have identical readings. Technician A says the coolant temperature sensor is faulty and should be replaced. Technician B says a failed ECM can cause this condition. Who is correct?

    A. A only

    B. B only

    C. Both A and B

    D. Neither A nor B

33. Refer to the composite vehicle (diagram on page 39 of the booklet) to answer this question: Technician A says the ECM receives unswitched battery power on Connector 3 ECM Pin 79. Technician B says the ECM receives switched battery power on Connector 3 ECM Pin 78. Who is correct?

    A. A only

    B. B only

    C. Both A and B

    D. Neither A nor B

34. Refer to the composite vehicle to answer this question: All of the following are acceptable voltage drop measurements on the ECM unswitched battery power wire Circuit 179 EXCEPT:

    A. 0.1 VDC.

    B. 0.01 VDC.

    C. 1.5 VDC.

    D. 0.15 VDC.

35. Refer to the composite vehicle for this question: Under what conditions is the high exhaust temperature sensor (HETS) lamp illuminated?

    A. The diesel particulate filter regeneration (DPFR) status lamp is illuminated.

    B. Exhaust gas temperature 1 (EGT1) is above 850°F (454.4°C) and vehicle speed is below 5 mph.

    C. EGT2 is above 850°F and vehicle speed is below 5 mph.

    D. EGT3 is above 850°F and vehicle speed is below 5 mph.

36. Refer to the composite vehicle to answer this question: The engine ECM has a DTC for an open coolant temperature sensor. The technician disconnects the wiring harness at connector F and jumpers across the two terminals. The ECM now sets a code for a shorted coolant temperature sensor. Which of the following could be the cause?

    A. A faulty ECM

    B. A faulty coolant temperature sensor

    C. A faulty wiring harness

    D. A faulty air intake temperature sensor

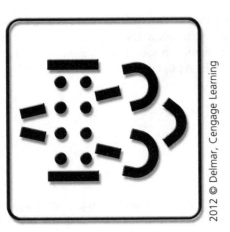

2012 © Delmar, Cengage Learning

37. Refer to the composite vehicle to answer this question: Referring to the figure above, the DPF status lamp is flashing on the dash of the composite vehicle. Which of the following is indicated?

A. The air intake filter is restricted.

B. The cabin air filter is restricted.

C. DPF filter regeneration is necessary because there is a moderate soot load.

D. The truck should be parked immediately and towed to a shop.

38. Refer to the composite vehicle to answer this question: At what temperature will the ECM derate the engine by 60 percent?

A. 225°F (107.2°C)

B. 230°F (110°C)

C. 235°F (112.8°C)

D. 240°F (115.6°C)

39. Refer to the composite vehicle to answer this question: Which of the following tools is LEAST LIKELY to be used when checking the data bus?

A. Ohmmeter

B. Voltmeter

C. Ammeter

D. Scan tool

40. Refer to the composite vehicle to answer this question: Cylinder #5 has a misfire. During diagnosis, the technician measures the resistance of the injector solenoid. Which of the following resistance readings indicates a good injector solenoid?

A. 0.4 ohms

B. 4.0 ohms

C. 0.2 ohms

D. 22.0 ohms

41. Refer to the composite vehicle to answer this question: The speedometer is displaying a faster vehicle speed than the vehicle is actually traveling. All of the following could cause this EXCEPT:

   A.   An incorrect rear axle ratio setting in the ECM.

   B.   An incorrect tire size rev/mile setting in the ECM.

   C.   An incorrect transmission tail shaft gear setting in the ECM.

   D.   An incorrect maximum cruise control setting in the ECM.

42. Refer to the composite vehicle for this question: Technician A says the idle engine speed programmable parameter can be set between 600 and 950 rpm. Technician B says the maximum power take-off (PTO) speed can be set between 500 and 850 rpm. Who is correct?

   A.   A only

   B.   B only

   C.   Both A and B

   D.   Neither A nor B

43. Refer to the composite vehicle to answer this question: Which of the following best describes the fuel system on the composite vehicle?

   A.   Common rail

   B.   HEUIs

   C.   PLN-E

   D.   EUIs

44. Refer to the composite vehicle to answer this question: Technician A says the engine oil pressure sensor provides the information for the dash gauge. Technician B says the accelerator pedal position (APP) sensor has three potentiometers. Who is correct?

   A.   A only

   B.   B only

   C.   Both A and B

   D.   Neither A nor B

45. Refer to the composite vehicle for this question: The stop engine light on the composite engine started flashing and 30 seconds later the engine shut down. Which of the following could be the cause?

   A.   A failed APP sensor

   B.   A failed after treatment DPF inlet temperature sensor (EGT2)

   C.   High engine oil temperature

   D.   Low engine coolant temperature

PREPARATION EXAM 2

1. All of the following can cause an immediate reduction in boost pressure EXCEPT:

   A. Leaking charge air cooler hoses.

   B. A leaking air filter gasket.

   C. A restricted air filter.

   D. A restricted diesel particulate filter (DPF).

2. A truck with an inactive DTC for EGT1 out of range is being repaired. After sitting overnight, a scan tool is connected to the engine. EGT1 indicates 80°F (26.7°C); EGT2 indicates 70°F (21.1°C); EGT3 indicates 75°F (23.9°C). Which of the following is the most likely cause of the code?

   A. EGT1 is faulty.

   B. EGT2 is faulty.

   C. EGT3 is faulty.

   D. A high-resistance electrical connection on EGT1.

3. Which of the following is LEAST LIKELY to cause a low power concern?

   A. A faulty ABS controller

   B. A restricted DPF

   C. A seized engine cooling fan clutch

   D. A faulty dash-mounted ECM diagnostic switch

4. Which of the following is LEAST LIKELY to be associated with a stuck open EGR valve?

   A. Excessive boost pressure

   B. Poor acceleration

   C. Excessive fuel pressure

   D. Low fuel pressure

5. A diesel engine dies after idling for three minutes. The engine will not restart. Technician A says low fuel level could be the cause. Technician B says the idle shutdown timer in the electronic control module (ECM) could be the cause. Who is correct?

   A. A only

   B. B only

   C. Both A and B

   D. Neither A nor B

6. Technician A says an ECM image can be electronically stored with the work order. Technician B says an ECM image can be saved to a file. Who is correct?

   A. A only

   B. B only

   C. Both A and B

   D. Neither A nor B

7. A DTC has been set for the variable valve actuator on a late-model diesel engine. Technician A says the variable valve actuator opens the exhaust valve. Technician B says the variable valve actuator opens the intake valve. Who is correct?

    A. A only

    B. B only

    C. Both A and B

    D. Neither A nor B

After treatment diesel oxidation catalyst (DOC) inlet temperature	700°F (371.1°C)
After treatment diesel particulate filter (DPF) inlet temperature	600°F (315.6°C)
After treatment diesel particulate filter (DPF) outlet temperature	500°F (260°C)

8. The readings shown above were taken 10 minutes after a stationary regeneration was started. Which of the following is indicated?

    A. A restricted DPF

    B. A damaged DOC

    C. A damaged DPF

    D. A damaged DPF outlet temperature sensor

9. All of the following could cause a diesel engine equipped with an electronic unit injector (EUI) fuel system to crank but not start EXCEPT:

    A. Low fuel level.

    B. A failed ECM.

    C. Low oil pressure.

    D. A failed engine position sensor.

10. During a stationary regeneration of the DPF, all of the following would normally occur EXCEPT:

    A. Turbo boost would increase.

    B. Exhaust temperature will rise.

    C. A DTC will be set for the DPF.

    D. Fuel will be injected into the exhaust stream.

11. Technician A says it may be necessary to loosen a bleeder screw when removing air from the fuel system. Technician B says it is standard procedure to crank the engine for four minutes at a time when bleeding the air from the fuel system. Who is correct?

    A. A only

    B. B only

    C. Both A and B

    D. Neither A nor B

12. Which of the following problems is LEAST LIKELY to set a DTC?

    A. An open after treatment injector solenoid

    B. A shorted after treatment injector solenoid

    C. An open engine brake on/off switch

    D. A shorted engine position sensor 2 (EPS2)

13. A vehicle has multiple counts of an inactive DTC for "coolant temperature sensor signal erratic." Which of the following is the most likely cause?

    A. Loose connector pins on the intake manifold temperature sensor

    B. Loose connector pins on the coolant temperature sensor

    C. A faulty intake manifold temperature sensor

    D. A faulty coolant temperature sensor

2012 © Delmar, Cengage Learning

14. Referring to the figure above, the lamp shown is illuminated on the dash. Which of the following is indicated?

    A. High exhaust restriction

    B. High exhaust temperature

    C. High intake restriction

    D. High intake temperature

15. The engine in a vehicle will not start. Technician A says an anti-theft security device could prevent the engine from starting. Technician B says the antilock brake system (ABS) controller could prevent the engine from starting. Who is correct?

    A. A only

    B. B only

    C. Both A and B

    D. Neither A nor B

16. Technician A says most ECMs will save freeze frame data on active codes. Technician B says most ECMs will save freeze frame data on inactive codes. Who is correct?

    A. A only

    B. B only

    C. Both A and B

    D. Neither A nor B

17. The technician has replaced a failed head gasket. After the test drive to confirm that the repair was successful, the ECM has an active trouble code for a damaged DPF. Which of the following is the most likely cause?

    A. A failed injector

    B. Damage to the DPF due to coolant from the failed head gasket

    C. A leaking charge air cooler

    D. A failed EGR valve

18. The remote power take-off (PTO) switch will not engage the remote PTO function. A technician has connected a scan tool to the engine and finds that the ECM indicates remote PTO "OFF" regardless of switch position. All of the following could be a cause EXCEPT:

    A. No voltage supplied to the PTO switch.

    B. A failed PTO drive gear.

    C. A failed PTO switch.

    D. A failed ECM.

19. The scan tool will not communicate with the engine. Which of the following is LEAST LIKELY cause?

    A. Low fuel level in the fuel tank

    B. No ground at the ECM

    C. No battery positive at the ECM

    D. A failed data bus

20. Technician A says the EGR valve helps reduce NO_x emissions. Technician B says the DPF helps reduce particulate emissions. Who is correct?

    A. A only

    B. B only

    C. Both A and B

    D. Neither A nor B

21. A diesel engine equipped with an EUI fuel system has a misfire on cylinder #2. The technician replaces the injector and the cylinder continues to misfire. Which of the following is the LEAST LIKELY cause?

    A. A worn cam lobe

    B. A failed injector harness

    C. A burned exhaust valve

    D. Low fuel supply pressure

22. The composite diesel engine has an inactive diagnostic trouble code (DTC) for high engine oil temperature. The freeze frame data indicates the engine coolant temperature was 160°F (71°C) and the intake manifold temperature was 100°F (37.8°C) when the condition occurred. Which of the following is the most likely cause of the stored DTC?

    A. A restricted radiator

    B. A restricted exhaust gas recirculation (EGR) cooler

    C. A faulty coolant temperature sensor

    D. A faulty oil temperature sensor

23. Refer to the composite vehicle to answer this question: Which of the following is the minimum speed at which a passive regeneration may occur?

    A. 10 mph

    B. 25 mph

    C. 30 mph

    D. 45 mph

24. Refer to the composite vehicle (diagram on page 39 of the booklet) answer this question: The engine has "low injector current" active codes for injectors 1 and 2. Technician A says an open circuit at ECM connector 1 Terminal 6 could be the cause. Technician B says an open engine ground at G30 could be the cause. Who is correct?

    A. A only

    B. B only

    C. Both A and B

    D. Neither A nor B

25. Refer to the composite vehicle (diagram on page 39 of the booklet) to answer the following question: While diagnosing an accelerator pedal position (APP) DTC using an oscilloscope, the technician finds a constant 5 volts at ECM connector 3, terminals 73 and 81, at all pedal positions. Which of the following could be the cause?

    A. Open on ECM terminal 72

    B. Short to ground at ECM terminal 72

    C. Failed ECM

    D. Failed APP sensor

26. Refer to the composite vehicle to answer this question: The ECM has set an active DTC for "crankcase pressure sensor signal shorted low." When Connector BX is disconnected, the code becomes inactive. Technician A says the wiring harness is open. Technician B says the sensor is shorted. Who is correct?

    A. A only

    B. B only

    C. Both A and B

    D. Neither A nor B

27. Refer to the composite vehicle to answer the following question: All of the following are true concerning the fuel system on the composite vehicle EXCEPT:

    A. The fuel transfer pump is mechanically driven.

    B. The engine uses an EUI fuel system with six unit injectors.

    C. All six injectors use a common power supply.

    D. The fuel temperature sensor is a potentiometer.

28. Refer to the composite vehicle to answer this question: While pulling a hill, the check engine light (CEL) and stop engine lamp (SEL) illuminate and the engine loses power. Technician A says engine coolant temperature above 230°F (110°C) could be the cause. Technician B says intake manifold temperature above 210°F (98.9°C) could be the cause. Who is correct?

    A. A only

    B. B only

    C. Both A and B

    D. Neither A nor B

29. Refer to the composite vehicle (diagram on page 40 of the booklet) to answer this question: The ambient air temperature sensor is multiplexed on which data bus?

    A. J1587

    B. J1708

    C. J1922

    D. J1939

30. Refer to the composite vehicle to answer this question: The composite vehicle has been in the shop multiple times for low power. All of the following could be the cause EXCEPT:

    A. High EGR exhaust gas temperature.

    B. High coolant temperature.

    C. High coolant pressure.

    D. High DPF restriction.

31. Refer to the composite vehicle to answer the following question: The truck will not accelerate above idle speed. Which of the following could be the cause?

    A. One APP sensor signal has failed.

    B. A failed engine oil pressure sensor.

    C. Low coolant level.

    D. A failed cruise control switch.

32. Refer to the composite vehicle to answer this question: A voltmeter is connected across terminals A and B of the clutch pedal switch. When the pedal is released, the voltmeter reads zero volts; when the pedal is depressed, the voltmeter reads 12 volts. Technician A says the switch is faulty. Technician B says the switch has an open ground. Who is correct?

    A. A only

    B. B only

    C. Both A and B

    D. Neither A nor B

33. Refer to the composite vehicle for the following question: The composite diesel engine has air bubbles in the cooling system. The head gasket was replaced. After the test drive, the cooling system again has air in it. Which of the following would be the LEAST LIKELY cause?

    A. Failure to fill the cooling system correctly

    B. A failed EGR cooler

    C. A cracked cylinder head

    D. High exhaust backpressure

34. Refer to the composite vehicle to answer the following question: The composite diesel engine has low power. Which of the following is LEAST LIKELY to be the cause?

    A. Cranking fuel pressure of 10 psi

    B. 80 psi fuel pressure at rated RPM

    C. Coolant temperature of 235°F (112.8°C)

    D. Engine oil temperature of 245°F (118.3°C)

35. Refer to the composite vehicle (diagram on page 39 of the booklet) to answer this question: Technician A says an open circuit 178 can cause a no-start condition. Technician B says an open circuit 171 can cause a no-start condition. Who is correct?

    A. A only

    B. B only

    C. Both A and B

    D. Neither A nor B

36. Refer to the composite vehicle to answer this question: Technician A says an engine cooling fan that runs all the time could be caused by a seized clutch. Technician B says an engine cooling fan that runs all the time could be caused by a faulty intake manifold pressure sensor. Who is correct?

    A. A only

    B. B only

    C. Both A and B

    D. Neither A nor B

37. Refer to the composite vehicle (diagram on page 38 of the booklet) to answer this question: The engine has a DTC for EGR valve non-communication. Which of the following could be the cause?

    A. A stuck open VGT

    B. A stuck closed EGR valve

    C. An open at connector JX pin 3

    D. A short at connector KX pin 1

38. Refer to the composite vehicle (diagram on page 37 of the booklet) to answer the following question: The composite vehicle has a misfire on cylinders #5 and #6. Any of the following could be the cause EXCEPT:

    A. A leaking head gasket.
    B. An open on splice S11.
    C. A loose injector hold down on cylinder #4.
    D. An open on splice S13.

39. Refer to the composite vehicle to answer this question: Which of the following engine oil pressure conditions would cause the engine to derate, but not shut down?

    A. 7 psi @ 1000 rpm
    B. 18 psi @ 1000 rpm
    C. 20 psi @ 1400 rpm
    D. 10 psi @ 1400 rpm

40. Refer to the composite vehicle (diagram on page 39 of the booklet) to answer this question: Neither the SEL nor the CEL will illuminate in any ignition switch position. The engine operates normally otherwise. Which of the following could be the cause?

    A. An open at ECM connector 3 pin 41
    B. An open at ECM connector 3 pin 42
    C. An open at splice S36
    D. An open circuit 148

41. Refer to the composite vehicle to answer the following question: Which of the following is LEAST LIKELY to cause the CEL to be illuminated?

    A. High coolant level
    B. Low coolant level
    C. Low oil level
    D. High fuel temperature

42. Refer to the composite vehicle (diagram on page 38 of the booklet) to answer this question: The EGR will not operate when commanded by the scan tool. The VGT works correctly. Which of the following could be the cause?

    A. An open circuit 314
    B. An open circuit 313
    C. An open circuit 320
    D. An open circuit 315

43. Refer to the composite vehicle to answer the following question: Technician A says the composite diesel engine is equipped with a returnless fuel system. Technician B says the fuel temperature sensor is located in the primary fuel filter. Who is correct?

    A. A only
    B. B only
    C. Both A and B
    D. Neither A nor B

44. Refer to the composite vehicle to answer this question: The engine cooling fan is running all the time. Technician A says a faulty intake manifold temperature sensor could be the cause. Technician B says a faulty coolant temperature sensor could be the cause. Who is correct?

    A. A only
    B. B only
    C. Both A and B
    D. Neither A nor B

45. Refer to the composite vehicle to answer the following question: The composite diesel engine will not start. There is no smoke from the exhaust while cranking. Technician A says this indicates no fuel is being delivered to the engine. Technician B says a faulty engine position sensor could be the cause of the no-start. Who is correct?

    A. A only
    B. B only
    C. Both A and B
    D. Neither A nor B

PREPARATION EXAM 3

1. An engine is diagnosed with low intake manifold pressure. The technician finds the turbocharger impeller severely worn. Which of the following could be the cause?

    A. A damaged air filter

    B. A stuck closed exhaust gas recirculation (EGR) valve

    C. A stuck open wastegate

    D. A failed after treatment fuel injector

2. A diesel engine will not start. The technician finds an active diagnostic trouble code (DTC) for glow plugs 1, 3, 5, and 7. Which of the following is the LEAST LIKELY cause of the no-start condition?

    A. A faulty ether injection solenoid valve

    B. An open glow plug wiring harness

    C. Low compression

    D. A failed glow plug controller

3. All of the following can cause a melted piston EXCEPT:

    A. A broken exhaust rocker.

    B. A stuck open injector.

    C. A stuck closed injector.

    D. A misdirected piston cooling nozzle.

4. A truck equipped with an automated manual transmission will not go into gear. Technician A says a failed clutch switch could be the cause. Technician B says a failed cruise control switch could be the cause. Who is correct?

    A. A only

    B. B only

    C. Both A and B

    D. Neither A nor B

5. Technician A says a loose exhaust manifold can cause low intake manifold pressure. Technician B says a loose intake manifold can cause low intake manifold pressure. Who is correct?

    A. A only

    B. B only

    C. Both A and B

    D. Neither A nor B

6. Technician A says that a faulty APP can cause a no-start. Technician B says that an ambient air temperature sensor can cause a no-start. Who is correct?

    A. A only

    B. B only

    C. Both A and B

    D. Neither A nor B

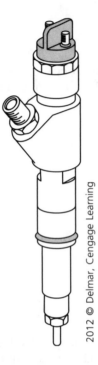

2012 © Delmar, Cengage Learning

7. Referring to the figure above, the injector shown is being replaced. Which of the following is LEAST LIKELY to be performed as part of the repair?

    A. The injector trim code may need to be installed in the ECM.

    B. The injector rocker must be adjusted.

    C. DTCs may need to be cleared from the ECM.

    D. The engine should have a test run after injector replacement.

8. A diesel engine is making a knocking noise. The technician performs a cylinder cutout test using the scan tool. When cylinder #4 is deactivated, the knocking stops. Technician A says a failed vibration damper could be the cause. Technician B says a failed dual mass flywheel could be the cause. Who is correct?

    A. A only

    B. B only

    C. Both A and B

    D. Neither A nor B

9. Technician A says the variable valve actuator system uses brake system air pressure to operate. Technician B says the variable valve actuator system only operates during engine brake operation. Who is correct?

   A. A only

   B. B only

   C. Both A and B

   D. Neither A nor B

10. The driver complains that the cruise control and speedometer do not operate on the truck. The antilock braking system (ABS) works correctly. Which of the following could be the cause?

    A. An open J1939 data bus

    B. A shorted J1939 data bus

    C. An open vehicle speed sensor (VSS)

    D. A shorted cruise control switch

11. An engine has had the turbocharger replaced due to a low boost pressure complaint. During the test drive after repairs, the technician finds the engine still has low boost. Which of the following is the LEAST LIKELY cause of the concern?

    A. A restricted fuel filter

    B. A restricted air filter

    C. A restricted DPF

    D. A restricted oil drain line

12. The vehicle has the "OPT IDLE" lamp on the dash illuminated. Which of the following is indicated?

    A. The vehicle will idle at a lower-than-normal RPM.

    B. The remote throttle is engaged.

    C. The APP has set a DTC and the vehicle will only idle.

    D. The truck will shut off and restart automatically.

13. The EGR system has an active DTC for the EGR airflow control (throttle) valve. Which of the following is the LEAST LIKELY cause of the code?

    A. A binding throttle valve

    B. A sticking throttle position sensor (TPS)

    C. A sticking APP sensor

    D. A shorted throttle valve wiring harness

14. Technician A says the cruise control switches can sometimes be used to change idle RPM. Technician B says the cruise control switches can be used to adjust maximum RPM. Who is correct?

    A. A only

    B. B only

    C. Both A and B

    D. Neither A nor B

15. A vehicle equipped with a hydraulically actuated electronic unit injector (HEUI) fuel system idles poorly and has low power. Any of the following could be the cause EXCEPT:

    A. The engine oil needs to be changed.

    B. The injectors are worn.

    C. The injector control pressure (ICP) sensor is faulty.

    D. The turbo charger wastegate is stuck open.

16. Technician A says the higher the resistance in an electrical circuit, the higher the amperage. Technician B says if voltage increases and resistance stays the same, the amperage will increase. Who is correct?

    A. A only

    B. B only

    C. Both A and B

    D. Neither A nor B

17. Refer to the composite vehicle (diagram on page 38 of the booklet) to answer this question: All of the following can cause a crankcase ventilation system DTC EXCEPT:

    A. An open circuit 302.

    B. An open circuit 305.

    C. A restricted crankcase filter.

    D. A missing crankcase filter.

18. Refer to the composite vehicle engine to answer this question: The composite vehicle engine has set a DTC for the EGR pressure differential sensor. The technician replaces the sensor and clears the DTC. After the test drive, an active DTC for the EGR pressure differential sensor has reappeared. Which of the following is the most likely cause?

    A. A faulty EGR pressure differential sensor

    B. Restricted sensor hoses

    C. A failed DPF

    D. A failed DOC

19. Refer to the composite vehicle (diagram on page 37 of the booklet) to answer this question: The engine brake is totally inoperative. Which of the following could be the cause?

    A. An open circuit at ECM connector 1 pin 9

    B. An open circuit at ECM connecter 1 pin 10

    C. An open circuit at ECM connector 1 pin 11

    D. An open circuit at ECM connector 1 pin 12

20. Refer to the composite vehicle to answer this question: The check engine lamp (CEL) and stop engine lamp (SEL) illuminate and the engine shuts off 30 seconds later. The technician connects a scan tool and finds low oil pressure indicated in the data stream. A master gauge is connected and obtains an engine reading of 20 psi @ 600 rpm. The oil pressure rises as the engine RPM is increased and reaches 40 psi @ 2000 rpm. Which of the following could be the cause of the shutdown concern?

    A. Worn main bearings

    B. A worn oil pump

    C. A faulty sending unit

    D. A stuck open pressure relief valve

21. Refer to the composite vehicle to answer this question: Technician A says the CEL will start flashing 30 seconds prior to idle shutdown. Technician B says the operator can override the idle shutdown timer by depressing the service brake pedal. Who is correct?

    A. A only
    B. B only
    C. Both A and B
    D. Neither A nor B

22. Refer to the composite vehicle to answer this question: Technician A says a faulty intake manifold temperature sensor can cause the engine cooling fan to stay on. Technician B says a faulty engine position sensor can cause the engine cooling fan to stay on. Who is correct?

    A. A only
    B. B only
    C. Both A and B
    D. Neither A nor B

23. Refer to the composite vehicle (diagram on page 40 of the booklet) to answer this question: There is an open circuit at connector SX pin 6. Which of the following would most likely be the result?

    A. The ECM would start performing an active DPF regeneration.
    B. The HETS lamp would start flashing.
    C. The ECM would start performing a passive regeneration.
    D. The SEL would illuminate.

24. Refer to the composite vehicle (diagram on page 39 of the booklet) to answer this question: Power take-off (PTO) mode works correctly from the remote PTO switch, but will not work from the cab-mounted PTO switch. Any of the following could be the cause EXCEPT:

    A. An open circuit 170.
    B. An EMC connector 3 pin 71 open.
    C. A failed PTO switch.
    D. An open circuit 171.

25. Refer to the composite vehicle to answer this question: The scan tool will communicate with the engine ECM, but will not communicate with the other modules on the truck. Which of the following could be the cause?

    A. An open circuit at connector XX
    B. An open circuit 562
    C. An open circuit 561
    D. An open circuit at connector YX terminal B

26. Refer to the composite vehicle to answer this question: Injector #3 was replaced due to rough idle and poor performance during a cylinder power balance test. During the test drive, the technician finds the problem is still there. Which of the following is the LEAST LIKELY cause?

   A.   Failure to reprogram the ECM with the new injector trim codes

   B.   Failure to set injector height correctly

   C.   A restricted primary fuel filter

   D.   A worn injector cam lobe

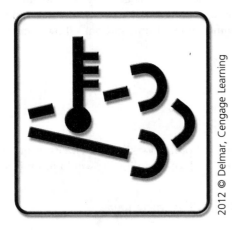

2012 © Delmar, Cengage Learning

27. Refer to the composite vehicle engine to answer this question: The lamp shown above is flashing. What does this indicate?

   A.   The exhaust gas temperature (EGT) 3 has indicated a temperature above 850°F (454.4°C).

   B.   The DPF delta pressure is higher than normal.

   C.   The coolant temperature is above 245°F (118.3°C).

   D.   The ECM has stored inactive codes.

28. Refer to the composite vehicle to answer this question: Which of the following is the correct fuel injector resistance specification?

   A.   0.05–0.5 ohms resistance

   B.   0.5–5.0 ohms resistance

   C.   5.0–50.0 ohms resistance

   D.   50.0–500.0 ohms resistance

29. Refer to the composite vehicle to answer this question: A driver is concerned about low power and poor fuel economy. Technician A says a shorted engine cooling fan control switch will cause the engine cooling fan to operate continuously. Technician B says an engine cooling fan solenoid with excessive resistance can cause the engine cooling fan to operate continuously. Who is correct?

   A.   A only

   B.   B only

   C.   Both A and B

   D.   Neither A nor B

30. Refer to the composite vehicle to answer this question: Technician A says the vehicle speed sensor should have a resistance of between 100 and 1000 ohms. Technician B says the vehicle speed sensor is a three wire sensor. Who is correct?

    A. A only
    B. B only
    C. Both A and B
    D. Neither A nor B

31. Refer to the composite vehicle to answer this question: Technician A says the coolant level sensor shares a ground with other switches. Technician B says the coolant level sensor connector is connector K. Who is correct?

    A. A only
    B. B only
    C. Both A and B
    D. Neither A nor B

32. Refer to the composite vehicle to answer this question: The technician is measuring the J1939 backbone resistance with both resistors in place. Which of the following would indicate a proper resistance?

    A. 60 ohms
    B. 120 ohms
    C. 240 ohms
    D. 360 ohms

33. Refer to the composite vehicle (diagram on page 40 of the booklet) to answer the following question: The technician is troubleshooting a no communication on J1939 condition. Technician A says one of the J1939 backbone resistors is located in the variable geometry turbocharger (VGT) actuator. Technician B says the J1939 circuit is connected at the ECM at connector XX, and at terminals A and B. Who is correct?

    A. A only
    B. B only
    C. Both A and B
    D. Neither A nor B

34. Refer to the composite vehicle to answer this question: Technician A says a restricted after treatment fuel injector can cause misfire on cylinder #5. Technician B says a restricted after treatment fuel injector can cause a low fuel system pressure.

    A. A only
    B. B only
    C. Both A and B
    D. Neither A nor B

35. Refer to the composite vehicle to answer this question: Technician A says active codes are indicated by the stop engine lamp (SEL) being illuminated. Technician B says inactive codes are indicated by the check engine light (CEL) being illuminated. Who is correct?

    A. A only

    B. B only

    C. Both A and B

    D. Neither A nor B

36. Refer to the composite vehicle engine to answer this question: All of the following can cause the composite vehicle engine to shut off EXCEPT:

    A. An idle shutdown timer.

    B. High coolant temperature.

    C. A restricted after treatment injector.

    D. Low oil pressure.

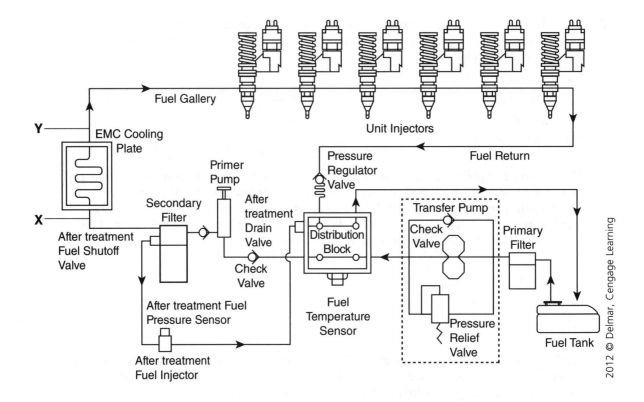

37. Refer to the composite vehicle engine to answer this question: Referring to the figure above, the composite vehicle engine has low power. With the engine at rated speed, Gauge X indicates 45 psi and Gauge Y indicates 44 psi. Which of the following is the LEAST LIKELY cause of low power?

    A. A restricted ECM cooling plate

    B. A restricted secondary filter

    C. A worn transfer pump

    D. A stuck open pressure regulator valve

38. Refer to the composite vehicle (diagram on page 39 of the booklet) to answer this question: The technician is diagnosing an APP1 DTC. While checking voltage at ECM connector 3 Terminal 73, the technician finds 0.5 volts at 0 percent depressed throttle and 1.0 volts at 100 percent depressed throttle. Which of the following could be the cause of the DTC?

    A. Circuit 172 is open.
    B. Circuit 173 is open.
    C. APP1 has failed.
    D. APP2 has failed.

39. Refer to the composite vehicle to answer this question: The engine will not start. The technician finds 0.1 volts at ECM connector 3, terminal 80, key-on engine-off. While cranking, the voltage at ECM connector 3, terminal 80 rises to 5.5 volts. Which of the following is the LEAST LIKELY source of the no-start?

    A. Circuit 180 has voltage drop.
    B. G30 is loose.
    C. S50 has corrosion.
    D. Circuit 179 has voltage drop.

40. Refer to the composite vehicle to answer this question: Which of the following is an acceptable reading from the intake manifold pressure sensor?

    A. 4.0 volts at 50 psig
    B. 3.3 volts at 37.5 psig
    C. 2.5 volts at 25 psig
    D. 1.5 volts at atmospheric pressure

41. Refer to the composite vehicle to answer this question: The composite diesel engine will not start. All of the following can cause the no-start EXCEPT:

    A. ECM supply voltage of 8.0 volts.
    B. ECM supply voltage of 17.0 volts.
    C. Fuel injector resistance of 30.0 ohms each.
    D. Fuel injector supply voltage of 120 volts.

42. Refer to the composite vehicle to answer this question: Technician A says a fuel temperature sensor signal of 275°F (135°C) could indicate a sensor with high resistance. Technician B says that the fuel system is used to cool the engine ECM. Who is correct?

    A. A only
    B. B only
    C. Both A and B
    D. Neither A nor B

43. Refer to the composite vehicle to answer this question: Which of the following is an acceptable intake temperature sensor resistance?

    A. 6–160 ohms
    B. 600–16,000 ohms
    C. 600–1600 ohms
    D. 6000–16,000 ohms

44. Refer to the composite vehicle to answer this question: Which of the following is LEAST LIKELY to cause a low power complaint?

    A. An open circuit 154
    B. An open circuit 16
    C. 45 psi fuel pressure at rated speed
    D. A faulty EGR Delta P sensor

45. Refer to the composite vehicle engine to answer this question: The driver complains that the idle shutdown timer will not function. Any of the following could be the problem EXCEPT:

    A. There is an active vehicle speed sensor fault.
    B. Ambient air temperature is 35°F (1.7°C).
    C. Engine coolant is 135°F (57.2°C).
    D. The service brake pedal switch is closed.

PREPARATION EXAM 4

1. A diesel engine cranks slowly and will not start. Technician A says low battery voltage may be the cause. Technician B says excessive voltage drop in the positive battery cables could be the cause. Who is correct?

    A. A only
    B. B only
    C. Both A and B
    D. Neither A nor B

Cylinder 1	15 amps
Cylinder 2	15 amps
Cylinder 3	0 amp
Cylinder 4	0 amp
Cylinder 5	0 amp
Cylinder 6	0 amp
Cylinder 7	0 amp
Cylinder 8	0 amp

2. Referring to the table of diagnostic measurements above, a diesel engine equipped with glow plugs will not start in cold weather. A scan tool and amp clamp were used to measure individual glow plug amperage. Which of the following is the most likely cause of the hard to start condition?

    A. An open glow plug timer
    B. Open glow plugs
    C. Low compression
    D. A faulty engine position sensor

3. The technician is uploading a new calibration file to the engine ECM while the ECM is still mounted on the engine. Which of the following would normally be part of this repair?

    A. Remove the negative battery cable during the reflash.
    B. Remove the positive battery cable during the reflash.
    C. Connect a battery maintainer (low-rate charger) on the battery during the reflash.
    D. Connect a high-rate charger on the battery during the reflash.

4. Which of the following is LEAST LIKELY to cause a diesel engine to fail to start?

    A. Low fuel pressure
    B. Low fuel level
    C. An open engine position sensor
    D. An open vehicle speed sensor (VSS)

5. The swinging vanes on a VGT are stuck. Which of the following will be the LEAST LIKELY result?

    A. An active non-communication DTC for the VGT

    B. An active position sensor DTC for the VGT

    C. Reduced horsepower

    D. Poor acceleration

6. The technician is performing a stationary regeneration of the DPF. The following temperatures are recorded 15 minutes after the regeneration is started: EGT1 = 660°F (348.9°C), EGT2 = 900°F (482.2°C), EGT3 = 975°F (523.9°C). Which of the following is true?

    A. EGT1 is higher than normal.

    B. EGT2 is higher than normal.

    C. All EGTs are normal.

    D. EGT3 is higher than normal.

7. Technician A says exhaust backpressure on trucks equipped with an exhaust after treatment device should be checked using a water manometer. Technician B says exhaust backpressure higher than normal can be caused by a missing DPF. Who is correct?

    A. A only

    B. B only

    C. Both A and B

    D. Neither A nor B

8. A diesel engine equipped with a hydraulically actuated electronically controlled unit injector (HEUI) fuel system will not start. Technician A says a faulty injection pressure regulator (IPR) can be the cause. Technician B says an open intake manifold pressure sensor can be the cause. Who is correct?

    A. A only

    B. B only

    C. Both A and B

    D. Neither A nor B

9. The driver complains that neither the cruise control nor the idle shutdown timer will work on a vehicle. Which of the following is the LEAST LIKELY cause?

    A. A sticking treadle valve

    B. A faulty VSS

    C. A faulty accelerator pedal position (APP) sensor

    D. A stuck cruise control switch

10. Technician A says a stationary regeneration of the DPF can be performed using a scan tool. Technician B says a stationary regeneration of the DPF can be performed using the dash-mounted switches. Who is correct?

    A. A only

    B. B only

    C. Both A and B

    D. Neither A nor B

11. The compression brake is weak on a diesel engine. The technician measures voltage at the engine brake solenoids and finds it to be within specification on all three. Which of the following is the LEAST LIKELY cause of the poor compression brake performance?

    A. Incorrect compression brake adjustment

    B. An open compression brake solenoid

    C. A leaking master piston seal

    D. An open brake selector switch

12. The cooling system will not hold pressure during a cooling system pressure test. Which of the following could be the cause?

    A. A leaking ECM cooler

    B. A leaking EGR cooler

    C. A leaking charge air cooler

    D. A leaking power steering cooler

13. Technician A says a damaged DPF can cause low power. Technician B says a damaged DOC can cause low power. Who is correct?

    A. A only

    B. B only

    C. Both A and B

    D. Neither A nor B

14. A diesel engine will not start. The technician disconnects the 5-volt reference wire for the pressure and temperature sensors from the ECM and the engine starts. Technician A says the ECM is faulty. Technician B says there is a short in the 5-volt reference circuit. Who is correct?

    A. A only

    B. B only

    C. Both A and B

    D. Neither A nor B

15. A diesel engine will not start. Which of the following could be the cause?

    A. A faulty DPF differential pressure sensor

    B. A failed engine position sensor

    C. A stuck closed after treatment injector

    D. A stuck closed after treatment fuel shutoff valve

16. The ECM has active DTCs for the engine oil temperature sensor, the ambient air temperature sensor, and the intake manifold temperature sensor. The freeze frame data indicates that the DTCs were all set at the same time. Which of the following could be the cause?

    A. An open engine oil temperature sensor

    B. An open common 5-volt reference wire

    C. An open signal wire from the intake manifold temperature sensor

    D. An overheated vehicle

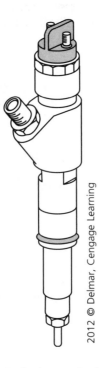

2012 © Delmar, Cengage Learning

17. Referring to the figure above, Technician A says the injector shown uses engine oil to create high-pressure fuel injection. Technician B says the injector shown above uses a camshaft lobe and rocker arm to create high pressure. Who is correct?

    A. A only

    B. B only

    C. Both A and B

    D. Neither A nor B

18. Refer to the composite vehicle to answer this question: The customer complains that the speedometer reads incorrectly. Any of the following could be the cause EXCEPT:

    A. Tire revolutions per mile set at 600.

    B. Tail shaft teeth set at 18.

    C. Rear axle ratio set at 5:1.

    D. Maximum engine speed without vehicle speed sensor (VSS) set at 2000 rpm.

19. Refer to the composite engine to answer this question: The ECM has set a diagnostic trouble code (DTC) for a nonresponsive VGT actuator motor. Which of the following is the LEAST LIKELY cause?

    A. A binding VGT actuator arm

    B. A sticking VGT nozzle ring

    C. Sticking VGT swinging vanes

    D. A stuck VGT motor

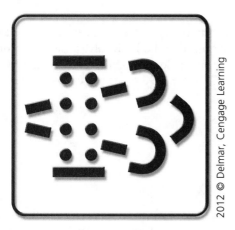

20. Refer to the composite engine to answer this question: The SEL and the light shown above are illuminated on the dash. What is indicated?

    A. Level 1 soot load

    B. Level 2 soot load

    C. Level 3 soot load

    D. Level 4 soot load

21. Refer to the composite engine to answer this question: All of the following could cause a misfire on cylinder #3 of the composite diesel engine EXCEPT:

    A. A restricted ECM cooling plate.

    B. A broken injector spring.

    C. A worn intake cam lobe.

    D. A worn exhaust cam lobe.

22. Refer to the composite vehicle to answer this question: Technician A says repeat head gasket failures may be caused by a restricted cooling system. Technician B says repeat head gasket failures can be caused by a restricted primary fuel filter. Who is correct?

    A. A only

    B. B only

    C. Both A and B

    D. Neither A nor B

23. Refer to the composite vehicle to answer this question: The diesel engine will not perform a stationary regeneration of the DPF. Technician A says a restricted EGR passage could be the cause. Technician B says a faulty after treatment injector could be the cause. Who is correct?

    A. A only
    B. B only
    C. Both A and B
    D. Neither A nor B

24. Refer to the composite vehicle to answer this question: The vehicle has set a DTC for EGR valve position. During diagnosis, the technician finds that the EGR valve position sensor voltage is 2.5 volts, regardless of engine RPM or engine load. Technician A says the EGR valve is stuck fully shut. Technician B says the EGR valve J1939 data bus circuit is open. Who is correct?

    A. A only
    B. B only
    C. Both A and B
    D. Neither A nor B

25. Refer to the composite vehicle to answer this question: The composite vehicle is being used in a refuse operation. The technician finds the diesel particulate filter (DPF) must have a stationary regeneration performed using a scan tool every two days. Any of following could be the cause EXCEPT:

    A. Mobile/active regeneration has been disabled.
    B. The high exhaust temperature lamp has been disabled.
    C. The DPF regeneration permit switch has been disabled.
    D. Stationary regeneration in power take-off (PTO) mode has been disabled.

26. Refer to the composite vehicle to answer this question: The engine cooling fan will not engage using the dash-mounted switch. It works correctly when commanded by the scan tool. Technician A says the switch may be stuck closed. Technician B says a failed engine cooling fan solenoid could be the cause. Who is correct?

    A. A only
    B. B only
    C. Both A and B
    D. Neither A nor B

27. Refer to the composite vehicle (diagram on page 37 of the booklet) to answer this question: Any of the following could cause the engine brake to function poorly EXCEPT:

    A. An open circuit at ECM connector 1 pin 9.
    B. An open circuit at ECM connecter 1 pin 10.
    C. An open circuit at ECM connector 1 pin 11.
    D. An open circuit at ECM connector 1 pin 12.

Cylinder 1	100 rpm
Cylinder 2	100 rpm
Cylinder 3	10 rpm
Cylinder 4	10 rpm
Cylinder 5	100 rpm
Cylinder 6	100 rpm

28. Refer to the composite vehicle to answer this question: The test results in the table above are from a power balance test on the composite engine. Which of the following is the most likely cause?

    A. A leaking cylinder head gasket

    B. A leaking EGR cooler gasket

    C. An open at splice S11

    D. An open at splice S13

29. Refer to the composite vehicle to answer this question: The diagnostic switch has been pressed. The SEL and CEL both momentarily turn on and then off. Which of the following could be the cause?

    A. There are multiple active fault codes.

    B. There are multiple inactive fault codes.

    C. There are no fault codes.

    D. The ECM is malfunctioning.

30. Refer to the composite vehicle to answer this question: All of the following are programmable parameters on the composite vehicle EXCEPT:

    A. Idle shutdown manual override.

    B. Minimum fan on-time.

    C. Tire size.

    D. Steering wheel diameter.

31. Refer to the composite vehicle to answer this question: Technician A says the ambient air temperature sensor is directly wired to the engine ECM. Technician B says the ambient air temperature sensor voltage should be 3.0 volts at 90°F (32.2°C). Who is correct?

    A. A only

    B. B only

    C. Both A and B

    D. Neither A nor B

32. Refer to the composite vehicle to answer this question: Technician A says a failed coolant temperature sensor can prevent the regeneration of the DPF. Technician B says a failed coolant temperature sensor can cause the instrument panel temperature gauge to read incorrectly. Who is correct?

    A. A only
    B. B only
    C. Both A and B
    D. Neither A nor B

33. Refer to the composite vehicle to answer this question: The fan control A/C pressure switch programmable parameter has been set to disabled. Which of the following is LEAST LIKELY to occur?

    A. The A/C will operate poorly at low vehicle speeds.
    B. The A/C will operate poorly at highway speeds.
    C. The A/C high-side pressure will be higher than normal.
    D. The A/C pressure switch will open.

34. Refer to the composite vehicle (diagram on page 40 of the booklet) to answer this question: The ECM has set an active DTC for EGT1 temperature below normal. Which of the following could be the cause?

    A. An open connector SX
    B. An open circuit 555
    C. An open circuit at ECM pin 456
    D. A short-to-ground on circuit 556

Fuel temperature sensor connected	0.0 volts
Fuel temperature sensor disconnected and shorted	5.0 volts

35. Refer to the composite vehicle to answer this question: There is an active DTC for the fuel temperature sensor. The technician connects the scan tool and monitors the fuel temperature sensor voltage while connecting and disconnecting the fuel temperature sensor. The test results are listed in the table above. Which of the following is the cause of the active DTC?

    A. The fuel temperature is higher than normal.
    B. The fuel temperature is lower than normal.
    C. The fuel temperature sensor is open.
    D. The fuel temperature sensor is shorted.

36. Refer to the composite vehicle to answer this question: All of the following are true concerning the exhaust gas temperature (EGT) sensors EXCEPT:

    A. EGT 1 is also referred to as the after treatment diesel oxidation catalyst (DOC) inlet temperature sensor.
    B. EGT 2 is also referred to as the after treatment DPF inlet temperature sensor.
    C. EGT3 is also referred to as the after treatment DPF outlet temperature sensor.
    D. An overheated EGT 1 will illuminate the high exhaust system temperature (HEST) lamp.

37. Refer to the composite vehicle to answer this question: The idle shutdown timer does not shut the vehicle off after the programmed time. Technician A says cold ambient temperatures could be the cause. Technician B says high ambient temperatures could be the cause. Who is correct?

    A. A only

    B. B only

    C. Both A and B

    D. Neither A nor B

Accelerator pedal position (APP)	0%
Engine speed	200 rpm
Camshaft position sensor/Engine position signal 1 (CMP/EPS1)	No
Crankshaft position sensor/Engine position signal 2 (CKP/EPS2)	Yes
Engine coolant level	Normal

38. Refer to the composite vehicle to answer this question: The composite diesel engine will not start. The scan tool data shown in the table above is retrieved while cranking the engine. Which of the following could be the cause?

    A. A faulty APP sensor

    B. A faulty EPS1

    C. A faulty EPS2

    D. Low coolant level

39. Refer to the composite engine (diagram on page 38 of the booklet) to answer this question: Technician A says wiring harness connector IX will be located on the exhaust gas recirculation (EGR) valve motor. Technician B says wiring harness connector JX will be a pigtail connector on the variable geometry turbocharge (VGT) actuator. Who is correct?

    A. A only

    B. B only

    C. Both A and B

    D. Neither A nor B

40. Refer to the composite vehicle to answer this question: Which of the following signals would be considered out of value range?

    A. After treatment DOC inlet temperature of 1000°F (537.8°C)

    B. After treatment DPF differential pressure of 28 in. Hg

    C. After treatment DPF outlet temperature of 1400°F (760°C)

    D. After treatment fuel pressure of 300 psi

41. Refer to the composite vehicle to answer this question: The engine oil pressure instrument panel gauge indicates low oil pressure. The scan tool indicates normal oil pressure. Technician A says the oil pressure sensor may be faulty. Technician B says there may be high resistance on ECM circuit 118. Who is correct?

    A. A only

    B. B only

    C. Both A and B

    D. Neither A nor B

42. Refer to the composite vehicle to answer this question: All of the following are true concerning the fuel system on the composite vehicle EXCEPT:

    A. The fuel transfer pump is mechanical.

    B. The injectors are fired in pairs.

    C. Injectors 1 and 2 share a common power supply.

    D. The system uses primary and secondary fuel filters.

43. Refer to the composite vehicle to answer this question: Which of the following would be LEAST LIKELY to cause the instrument panel temperature gauge to show higher than normal operating coolant temperature?

    A. Restricted coolant passages in the radiator

    B. Higher-than-normal resistance in the engine coolant temperature (ECT) sensor

    C. Restricted external passages through the charge air cooler

    D. A damaged water pump impeller

44. Refer to the composite vehicle (diagram on page 38 of the booklet) to answer this question: There is an active DTC set for exhaust backpressure out-of-range high. Any of the following could be the cause EXCEPT:

    A. A restricted DOC.

    B. A short between terminals 1 and 3 in connector AX.

    C. A short between terminal 1 and 2 in connector AX.

    D. A restricted DPF.

45. Refer to the composite vehicle to answer this question: The composite vehicle is hard to start after sitting overnight. What should the technician do to isolate the problem?

    A. Advise the driver to spray ether into the intake manifold to help start the engine.

    B. Use the hand primer pump to determine if the fuel system has lost prime.

    C. Remove the injectors and bench test them.

    D. Replace the ECM cooling plate.

PREPARATION EXAM 5

1. Technician A says a restricted air cleaner can cause slow acceleration. Technician B says a restricted diesel oxidation catalyst (DOC) can cause slow acceleration. Who is correct?

    A. A only
    B. B only
    C. Both A and B
    D. Neither A nor B

EGR temperature	70°F (21.1°C)
Coolant temperature	50°F (10°C)
Oil temperature	70°F (21.1°C)
Ambient air temperature	70°F (21.1°C)

2. Refer to the table above. The driver of a vehicle complains of poor fuel economy. After the vehicle has been sitting on the lot for 24 hours, the technician connects the scan tool and finds there are no active or inactive DTCs, then reviews the sensor values listed in the table. Technician A says the coolant is diluted and needs to be drained and refilled. Technician B says the ambient air temperature sensor is indicating an incorrect temperature and needs to be replaced. Who is correct?

    A. A only
    B. B only
    C. Both A and B
    D. Neither A nor B

3. Which of the following air inlet restriction test results indicate that the air filter needs to be changed?

    A. 5 in. H_2O
    B. 1 in. Hg
    C. 25 in. H_2O
    D. 30 in. Hg

4. Technician A says if an image is taken of the ECM prior to performing repairs, the Technician will have a before-and-after documentation of repairs. Technician B says downloading and saving ECM data before replacing the ECM is advisable. Who is correct?

    A. A only
    B. B only
    C. Both A and B
    D. Neither A nor B

5. The turbocharger was replaced due to a worn compressor wheel. The engine is still low on power. Which of the following is the LEAST LIKELY cause?

   A. The piston rings are worn.

   B. The primary fuel filter is restricted.

   C. The secondary fuel filter is restricted.

   D. The ECM is faulty.

After treatment diesel oxidation catalyst (DOC) inlet temperature	70°F (21.1°C)
After treatment diesel particulate filter (DPF) inlet temperature	600°F (315.5°C)
After treatment diesel particulate filter (DPF) outlet temperature	500°F (260°C)

6. Referring to the table above, the readings shown were taken 10 minutes after a stationary regeneration was started. Which of the following is indicated?

   A. A restricted DPF

   B. A damaged DOC

   C. A damaged DPF

   D. A damaged DOC inlet temperature sensor

7. A truck equipped with an HPCR fuel system has an active DTC that reads, "unable to achieve desired common rail pressure." Which of the following is the LEAST LIKELY cause?

   A. A restricted primary fuel filter

   B. A restricted high-pressure drain line overflow

   C. Worn injectors

   D. A restricted secondary fuel filter

8. A technician is testing the resistance of the glow plug. Any of the following would indicate an unacceptable resistance EXCEPT:

   A. 0.5 ohms.

   B. 5.0 ohms.

   C. 50.0 ohms.

   D. Out of limit (OL) ohms.

9. A diesel engine has had multiple EGR coolers replaced. Technician A says a leaking head gasket could be the cause. Technician B says a leaking cooling system could be the cause. Who is correct?

   A. A only

   B. B only

   C. Both A and B

   D. Neither A nor B

Engine RPM	275 rpm
HPCR fuel pressure	250 psi
Intake air temperature	75°F (23.9°C)

10. Referring to the table above, these scan tool readings were taken while cranking an engine equipped with a high-pressure common rail (HPCR) fuel system that will not start. Which of the following is the LEAST LIKELY cause of the no-start condition?

    A. A failed EPS

    B. A restricted primary fuel filter

    C. A restricted secondary fuel filter

    D. A failed HPCR pump

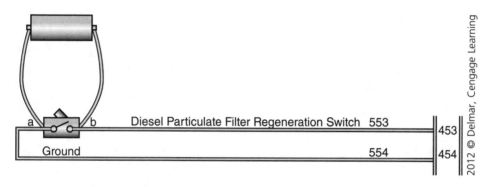

11. Referring to the figure above, when a digital volt-ohmmeter (DVOM) is connected, the DVOM meter reads 12 volts, regardless of switch position. Which of the following is the most likely cause?

    A. A stuck closed (shorted) switch

    B. A stuck open (open) switch

    C. An open in circuit 553

    D. An open in circuit 554

12. At hot idle, the stop engine light (SEL) began flashing and then the engine shut down. Any of the following could be the cause EXCEPT:

    A. The idle oil pressure reading shows 4 psi.

    B. The engine coolant temperature is 242°F (116.7°C).

    C. The engine oil temperature is 265°F (129.4°C).

    D. The coolant level is below normal level.

13. Technician A says a stuck open wastegate can cause excessive intake manifold pressure. Technician B says a stuck open wastegate can cause excessive exhaust backpressure. Who is correct?

    A. A only

    B. B only

    C. Both A and B

    D. Neither A nor B

14. Which of the following would LEAST LIKELY be used to diagnose a misfiring cylinder on a diesel engine equipped with an HPCR fuel system?

    A. A 1000 ml beaker

    B. An ohmmeter

    C. A height gauge

    D. A scan tool

15. Technician A says sticking swinging vanes in the VGT can cause an intermittent power loss. Technician B says a sticking EGR valve can cause intermittent power loss. Who is correct?

    A. A only

    B. B only

    C. Both A and B

    D. Neither A nor B

16. Technician A says a dial indicator can be used to check wastegate actuator travel. Technician B says the wastegate actuator arm must be removed from the turbocharger to measure its travel. Who is correct?

    A. A only

    B. B only

    C. Both A and B

    D. Neither A nor B

17. Technician A says excessive pressure in the HPCR can be caused by a faulty o-ring on the HPCR fuel system. Technician B says excessive pressure in the HPCR fuel system can be caused by a leaking high-pressure overflow valve. Who is correct?

    A. A only

    B. B only

    C. Both A and B

    D. Neither A nor B

Cylinder 1	15 amps
Cylinder 2	0 amp
Cylinder 3	15 amps
Cylinder 4	0 amp
Cylinder 5	15 amps
Cylinder 6	0 amp
Cylinder 7	15 amps
Cylinder 8	0 amp

18. Referring to the results from diagnostic measurements in the table above, a diesel engine equipped with glow plugs will not start in cold weather. The scan tool and an amp clamp are used to measure individual glow plug amperage. Which of the following is the most likely cause of the hard to start condition?

    A. An open glow plug timer

    B. An open glow plug harness

    C. Low compression

    D. A faulty engine position sensor

19. Refer to the composite vehicle to answer this question: The composite diesel engine has low power and has set an active DTC for low pressure at the after treatment fuel pressure sensor. Any of the following could be the cause EXCEPT:

    A. A restricted primary fuel filter.

    B. A restricted secondary fuel filter.

    C. A restriction in the distribution block.

    D. A restricted ECM cooling plate.

20. Refer to the composite vehicle to answer this question: The diesel particulate filter (DPF) requires regeneration more often than normal. Which of the following would be the most likely cause?

    A. The fan control switch failed open

    B. Use of the wrong diesel fuel

    C. An open engine brake solenoid winding

    D. An open at connector K Terminal 1

21. Refer to the composite vehicle to answer this question: The composite diesel engine has had the turbocharger replaced due to a separated turbine wheel. Now the engine lacks power. Technician A says the DOC may be restricted. Technician B says the electronic control module (ECM) should be checked for diagnostic trouble codes (DTCs). Who is correct?

    A. A only

    B. B only

    C. Both A and B

    D. Neither A nor B

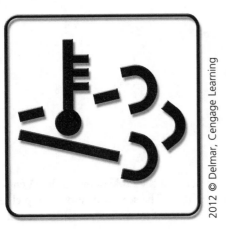

2012 © Delmar, Cengage Learning

22. Refer to the composite vehicle to answer this question: During a stationary regeneration, the lamp above is illuminated. Technician A says this indicates a high exhaust temperature on EGT3. Technician B says this indicates an overheat condition, and the stationary regeneration should be stopped. Who is correct?

A. A only

B. B only

C. Both A and B

D. Neither A nor B

23. Refer to the composite engine to answer this question: Technician A says that tire revolutions per mile is an adjustable parameter in the ECM. Technician B says the idle shutdown timer is an adjustable parameter in the ECM. Who is correct?

A. A only

B. B only

C. Both A and B

D. Neither A nor B

Cylinder 1	0 rpm
Cylinder 2	0 rpm
Cylinder 3	100 rpm
Cylinder 4	100 rpm
Cylinder 5	100 rpm
Cylinder 6	100 rpm

24. Refer to the composite vehicle to answer this question: The results from a power balance test on the composite engine are shown in the table above. Which of the following is the most likely cause for these readings?

A. An open solenoid on injector #1

B. A leaking charge air cooler gasket

C. An open circuit at splice S11

D. An open circuit at splice S13

25. Refer to the composite vehicle (diagram on page 39 of the booklet) to answer this question: Technician A says circuits 147 and 148 are a twisted pair. Technician B says the vehicle speed sensor (VSS) is connected directly to the J1939 data bus. Who is correct?

    A. A only
    B. B only
    C. Both A and B
    D. Neither A nor B

26. Refer to the composite vehicle to answer this question: Which of the following would LEAST LIKELY be used to locate a misfiring cylinder on the composite engine?

    A. An ohmmeter
    B. A scan tool
    C. A pyrometer
    D. A fuel pressure gauge

27. Refer to the composite vehicle to answer this question: The composite diesel engine will not start. The technician connects the scan tool and finds active DTCs for the following sensors: intake manifold temperature (IMT), intake manifold pressure, EGT1, and EPS1. Which of the following is the most likely cause of the no-start?

    A. Intake manifold temperature
    B. Intake manifold pressure
    C. EGT1
    D. Engine position sensor 1

28. Refer to the composite vehicle to answer this question: Technician A says fuel pressure should stay constant regardless of engine RPM. Technician B says excessive fuel pressure can be caused by a stuck open pressure regulator valve. Who is correct?

    A. A only
    B. B only
    C. Both A and B
    D. Neither A nor B

29. Refer to the composite engine to answer this question: The truck overheats and shuts down. Which of the following could be the cause?

    A. A shorted cooling fan control switch
    B. Circuit 158 shorted to battery positive
    C. An open engine cooling fan solenoid winding
    D. A leaking air line to the fan clutch

30. Refer to the composite vehicle (diagram on page 37 of the booklet) to answer this question: The composite diesel engine has active DTCs for the engine oil temperature (EOT) sensor and the inlet air temperature (IAT) sensor. Which of the following is the LEAST LIKELY cause?

    A. An open circuit at splice S32
    B. An open circuit at splice S31
    C. An open circuit at ECM connector 1 terminal 25
    D. An open in circuit 222

31. Refer to the composite vehicle to answer this question: All of the following are true concerning the affects of DPF soot loading EXCEPT:

    A. At full soot load the check engine light (CEL) is illuminated.
    B. Derate begins at full soot load.
    C. The DPF is overfull at 81 percent soot load.
    D. Power derate is 80 percent when the DPF is overfull.

32. Refer to the composite vehicle to answer this question: The after treatment fuel pressure sensor indicates fuel pressure varying with engine RPM from 50–90 psi any time the engine is running. Which of the following is the most likely cause?

    A. The sensor is faulty.
    B. The after treatment fuel shutoff valve is open.
    C. The after treatment fuel shutoff valve is closed.
    D. The after treatment drain valve is open.

33. Refer to the composite vehicle to answer this question: The after treatment DPF inlet temperature sensor needs to be replaced. Where is the sensor located on the vehicle?

    A. At the inlet of the DOC
    B. Between the DOC and DPF
    C. At the outlet of the DPF
    D. At the outlet of the turbocharger

Exhaust backpressure	110 in. Hg
Inlet air temperature	100°F (37.8°C)
Coolant temperature	180°F (82.2°C)
Intake manifold pressure	50 psi

34. Refer to the composite vehicle to answer this question: The composite diesel engine has a low power concern. The scan tool data shown in the table above was captured with the vehicle under a medium load at 55 mph. Which of the following could be the cause of the low power concern?

    A. Low intake manifold pressure
    B. High coolant temperature
    C. High intake air temperature
    D. High exhaust backpressure

Battery voltage	13.6 volts
Cranking RPM	95 rpm
Coolant temperature	25°F (−3.9°C)
Intake air temperature	23°F (−5°C)

35. Refer to the composite vehicle to answer this question: The vehicle is hard to start, especially on a cold morning. The measurements recorded in the table above were taken with the engine cranking. Which of the following could be the cause of the concern?

    A. Low battery voltage
    B. Key-off engine-off battery drain
    C. High resistance in the battery cables
    D. Incorrect coolant temperature sensor data

36. Refer to the composite vehicle booklet to answer this question: The composite vehicle has an inactive DTC stored for high fuel temperature. Which of the following is the LEAST LIKELY cause of the inactive code?

    A. Low fuel level in the tank
    B. An intermittent short at connector H
    C. High resistance at ECM connector 1 terminal 20
    D. High fuel system temperature occurred

37. Refer to the composite vehicle to answer this question: The composite engine has an active DTC for the fuel temperature sensor. The technician checks voltage at connector H. With the connector connected to the sensor, the voltmeter reads 0.0 volts. With the connector disconnected from the sensor, the voltmeter reads 5.0 volts. When the same test is performed at ECM connector 1 Pins 21 and 22, the voltmeter reads 5.0 volts with the sensor connected or disconnected. Technician A says the ECM has failed and must be replaced. Technician B says the sensor has failed and must be replaced. Who is correct?

    A. A only
    B. B only
    C. Both A and B
    D. Neither A nor B

38. Refer to the composite vehicle (diagram on page 40 of the booklet) to answer this question: An ohmmeter is connected at ECM connector 4, terminals 464 and 465. With the key off, the ohmmeter reads out of limit (OL or infinity) ohms. All of the following could be the cause EXCEPT:

    A. An open in one of the J1939 backbone resistors.
    B. An open in circuit 566 between the ECM and connector XX.
    C. An open in circuit 567 between the ECM and connector XX.
    D. Connector XX is unplugged.

39. Refer to the composite vehicle to answer this question: The engine runs poorly, has an active DPF fault, and blows black smoke. Technician A says worn injectors could be the cause. Technician B says a worn transfer pump could be the cause. Who is correct?

    A. A only
    B. B only
    C. Both A and B
    D. Neither A nor B

40. Refer to the composite vehicle to answer this question: The ECM shows a DTC for a Level 3 soot load and the technician is preparing the vehicle to perform a stationary regeneration. Which of the following would be the maximum time the regeneration should take?

    A. 0.5 hours
    B. 1.0 hours
    C. 1.5 hours
    D. 2.0 hours

41. Refer to the composite vehicle to answer this question: The engine has an active DTC for the DPF differential pressure sensor (DPF Delta P). All of the following are true concerning the sensor EXCEPT:

    A. The sensor measures pressure drop.

    B. The sensor is mounted on the DPF.

    C. The sensor has two ports.

    D. One side of the sensor monitors pressure before the DOC and the other side of the sensor monitors pressure after the DPF.

42. Refer to the composite vehicle (diagram on page 39 of the booklet) to answer this question: The composite vehicle has an open circuit between battery positive and S54. Which of the following is LEAST LIKELY to occur?

    A. The engine will not start.

    B. The engine ECM will not communicate with the scan tool.

    C. The engine will idle poorly.

    D. The engine ECM will not communicate with other modules on J1939.

43. Refer to the composite vehicle to answer this question: The composite diesel engine has a DTC for "invalid EPS2 sensor signal." Technician A says this can cause the engine to fail to start. Technician B says this can cause the engine to shut down. Who is correct?

    A. A only

    B. B only

    C. Both A and B

    D. Neither A nor B

44. Refer to the composite vehicle (diagram on page 37 of the booklet) to answer this question: All of the following can cause a no-start condition on the composite diesel engine EXCEPT:

    A. An open connector C.

    B. An open connector A.

    C. Failed engine position sensor 1 (EPS1).

    D. Failed EPS2.

45. Refer to the composite vehicle to answer this question: The composite diesel engine runs poorly. The technician disconnects connector A and the engine continues to run. Which of the following could be the cause?

    A. A failed ECM

    B. Failed injectors

    C. A worn engine camshaft

    D. A damaged tone wheel for EPS1

PREPARATION EXAM 6

Engine RPM	275 rpm
HPCR fuel pressure	250 psi
Intake air temperature	75°F (23.9°C)

1. Referring to the table of readings above, an engine equipped with a high-pressure common rail (HPCR) fuel system will not start. The scan tool readings were taken while cranking the engine. Which of the following is the LEAST LIKELY cause of the no-start condition?

    A. A failed EPS

    B. Leaking injectors

    C. A leaking pressure relief valve

    D. A failed HPCR pump

2. The head gasket has been replaced twice on a vehicle in fewer than 5,000 miles. Technician A says the head bolts may be stretched. Technician B says incorrect injector calibration codes installed in the ECM could be the cause. Who is correct?

    A. A only

    B. B only

    C. Both A and B

    D. Neither A nor B

3. A vehicle with an electronic unit injector (EUI) fuel system has had all six injectors replaced due to a black smoke concern. Now the truck idles poorly. Technician A says the injector height may not have been set properly during the installation. Technician B says the injector calibration codes may not have been installed properly. Who is correct?

    A. A only

    B. B only

    C. Both A and B

    D. Neither A nor B

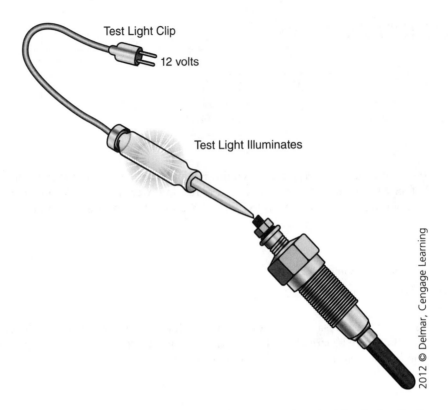

Test Light Clip

12 volts

Test Light Illuminates

2012 © Delmar, Cengage Learning

4. Referring to the figure above, a diesel engine equipped with a hydraulically actuated electronically controlled unit injector (HEUI) fuel system will not start. The technician performs the test above on all the glow plugs and the test light illuminates on every glow plug. Any of the following could be the cause of the no-start condition EXCEPT:

   A. Faulty glow plugs.
   B. Low oil level in the oil pan.
   C. A faulty glow plug controller.
   D. Low injection actuation pressure.

5. A vehicle that has had a turbocharger replacement now has low power and high exhaust temperature. Technician A says a portion of the old exhaust turbine may be lodged in the exhaust pipe. Technician B says the wastegate may be stuck open on the new turbocharger. Who is correct?

   A. A only
   B. B only
   C. Both A and B
   D. Neither A nor B

6. Technician A says a smoke machine can be used to test the integrity of the exhaust system. Technician B says a smoke machine can be used to check the integrity of the intake system. Who is correct?

   A. A only

   B. B only

   C. Both A and B

   D. Neither A nor B

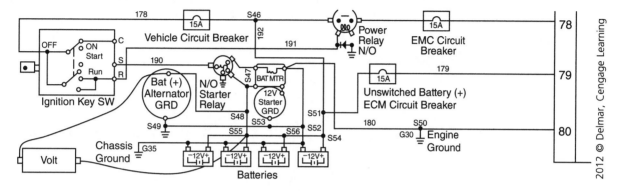

7. Referring to the figure above, a voltage drop test is performed on the composite engine. With the engine running at fast idle and all the electrical accessories on the vehicle turned on, the voltmeter indicates 3.6 VDC. Which of the following could be the cause?

   A. Low battery voltage

   B. Excessive current draw of electrical accessories

   C. A faulty alternator ground

   D. High resistance between S48 and S55

8. The diesel engine cranks slowly. Which of the following would LEAST LIKELY be used to test the batteries?

   A. A load tester

   B. A voltmeter

   C. An ohmmeter

   D. A capacitance tester

9. An engine has an active DTC for an accelerator pedal position (APP) sensor. The APP is a Hall-effect-style sensor. Which of the following tools would the technician most likely use to diagnose the sensor?

   A. Ohmmeter

   B. Ammeter

   C. Scan tool

   D. Oscilloscope

10. A diesel engine that is not equipped with a DPF cranks, but will not start. There is no smoke from the exhaust while cranking. Any of the following could be the cause EXCEPT:

    A. No voltage to the ECM.

    B. No fuel to the engine.

    C. Low compression.

    D. A failed engine position sensor.

11. A diesel engine will not start. A symbol of a key is flashing on the instrument panel. Which of the following is LEAST LIKELY to be the cause of the no-start?

    A. Low fuel level

    B. An incorrect password entered

    C. An incorrect key being used

    D. A failed vehicle security system

12. Which of the following would LEAST LIKELY be used to diagnose a misfiring cylinder on an HEUI fuel system?

    A. Scan tool

    B. Ohmmeter

    C. 1000 ml beaker

    D. Pyrometer

13. The technician takes a test drive in a vehicle after replacing a failed DPF Delta P sensor. The malfunction indicator lamp (MIL) is still on. The engine is heavy-duty onboard diagnostic (HD-OBD) compliant. The SEL and the CEL are both off. The technician connects the scan tool and the engine ECM indicates no active trouble codes. Which of the following is the most likely cause?

    A. A failed ECM

    B. A failed J1939 data bus

    C. A failed scan tool

    D. Technician failed to reset the MIL

14. An HPCR fuel system is being tested for fuel return flow from the overflow valve. Technician A says excessive fuel return can be caused by restricted fuel filters. Technician B says excessive fuel return can be caused by a restricted high-pressure relief valve. Who is correct?

    A. A only

    B. B only

    C. Both A and B

    D. Neither A nor B

15. The ECM has set a fault code for EPS1. During troubleshooting, the technician disconnects the EPS1 wiring harness and finds the internal terminals covered in engine oil. Which of the following is the most likely cause?

    A. High engine oil pressure

    B. High engine crankcase pressure

    C. A failed EPS1

    D. A leaking valve cover gasket

16. An engine with an HPCR fuel system starts okay when cold, but is hard to start when warm. The customer does not have any other concerns. Technician A says faulty injectors could be the cause. Technician B says a restricted air filter could be the cause. Who is correct?

    A. A only

    B. B only

    C. Both A and B

    D. Neither A nor B

17. Technician A says incorrect calibration files can cause poor fuel economy. Technician B says incorrect calibration files can cause poor engine performance. Who is correct?

    A. A only

    B. B only

    C. Both A and B

    D. Neither A nor B

18. An engine equipped with an HPCR fuel system has a DTC for "unable to reach desired fuel rail pressure." Technician A says the problem can be worn injectors. Technician B says the problem can be a worn high-pressure pump. Who is correct?

    A. A only

    B. B only

    C. Both A and B

    D. Neither A nor B

19. A diesel engine that is not equipped with a DPF blows smoke and is hard to start. Any of the following could be the cause EXCEPT:

    A. Worn compression rings.

    B. Worn injectors.

    C. A restricted engine air filter.

    D. A stuck open wastegate.

20. All of the following should be done prior to installing a new calibration file EXCEPT:

    A. Warm the engine to operating temperature.

    B. Ensure the batteries are fully charged.

    C. Connect the scan tool to an external power supply.

    D. Connect the scan tool to the engine ECM.

21. Refer to the composite vehicle to answer this question: The composite vehicle has a DTC for misfire on cylinder #2. During testing, cylinder #2 fails a cylinder contribution test. A compression test is performed and compression is found to be acceptable. Any of the following could be the cause of the DTC EXCEPT:

    A. A leaking exhaust valve.

    B. A worn injector cam lobe.

    C. A worn injector.

    D. High resistance in the injector solenoid.

22. Refer to the composite vehicle to answer this question: The operator complains of the engine stalling under a long pull. Technician A says the electronic control module (ECM) should be checked for freeze frame data to help diagnose the concern. Technician B says the vehicle may need to be connected to a loaded trailer and driven to duplicate the concern. Who is correct?

    A. A only

    B. B only

    C. Both A and B

    D. Neither A nor B

23. Refer to the composite vehicle to answer this question: Technician A says a shorted injector harness can cause a DTC. Technician B says an injector harness can be tested using an ohmmeter. Who is correct?

    A. A only

    B. B only

    C. Both A and B

    D. Neither A nor B

24. Refer to the composite vehicle to answer this question: The composite diesel engine has a DTC for high crankcase pressure. Which of the following is LEAST LIKELY to be the cause?

    A. A restricted engine air filter

    B. A restricted crankcase ventilation filter

    C. Worn piston rings

    D. A hole in a piston

25. Refer to the composite vehicle to answer this question: The composite engine has a knock that is evident at 800 rpm, 1200 rpm, and 2000 rpm. Which of the following is the most likely cause?

    A. A loose rod bearing

    B. A loose main bearing

    C. A loose piston pin

    D. A failed vibration damper

26. Refer to the composite vehicle to answer this question: The vehicle has a low power concern, but no DTCs. During diagnosis, intake manifold pressure at rated speed under full load is 25 psi. Technician A says the low power concern may be due to a missing diesel particulate filter (DPF). Technician B says the low power concern may be due to a restricted fuel filter. Who is correct?

    A.   A only

    B.   B only

    C.   Both A and B

    D.   Neither A nor B

27. Refer to the composite vehicle to answer this question: The vehicle has a low power complaint. The technician verifies the concern and finds an active DTC for low intake manifold pressure. Which of the following is the LEAST LIKELY cause?

    A.   A worn turbo impeller wheel

    B.   A worn turbo exhaust turbine

    C.   A restricted DPF

    D.   A malfunctioning theft deterrent system

28. Refer to the composite vehicle to answer this question: The customer has a low power concern. At full load, the technician finds 20 psi intake manifold pressure. There are no DTCs. Which of the following is the LEAST LIKELY cause of the concern?

    A.   Restricted fuel filters

    B.   A restricted air filter

    C.   An open on ECM pin 214

    D.   Worn impeller blades on the turbo

29. Refer to the composite vehicle (diagram on pages 37 and 40 of the booklet) to answer this question: The engine cranks but fails to start. Any of the following could be the cause EXCEPT:

    A.   An open on circuit 554.

    B.   An open circuit 223.

    C.   Connector A is open.

    D.   A short between ECM pins 27 and 28.

30. Refer to the composite diesel engine (diagram on pages 37 and 39 of the booklet) to answer this question: The engine brake works on medium and high, but it does not work on low. Which of the following could be the cause?

    A.   An open circuit 111

    B.   An open circuit 110

    C.   A failed engine brake on/off switch

    D.   An open on circuit 155

Engine coolant temperature	180°F (82.2°C)
Engine RPM	600 rpm
Engine oil pressure	1 psi
Exhaust backpressure	0.1 psi

31. Refer to the composite vehicle to answer this question: The composite vehicle's check engine light (CEL) and stop engine light (SEL) are illuminated and the engine shuts off while idling in traffic. The freeze frame data in the table above was retrieved from the ECM. Which of the following could be the cause of the engine shutdown?

    A. The engine was overheated.

    B. The oil level was too low.

    C. The engine was idling too fast.

    D. There was excessive exhaust backpressure.

32. Refer to the composite vehicle to answer this question: When the engine brake is engaged, the engine cooling fan does not engage. There are no trouble codes, and the driver has no other complaints. Technician A says incorrect programming could be the cause. Technician B says an open to the engine cooling fan solenoid could be the cause. Who is correct?

    A. A only

    B. B only

    C. Both A and B

    D. Neither A nor B

33. Refer to the composite vehicle to answer this question: Which of the following would LEAST LIKELY cause a diesel engine to start and die?

    A. An open in circuit 128

    B. Air in the fuel system

    C. A loose primary fuel filter

    D. Restriction in the fuel suction line

34. Refer to the composite vehicle to answer this question: Technician A says a restricted DPF could be caused by a restricted after treatment fuel shutoff valve. Technician B says a restricted DPF could be caused by an after treatment drain valve that has failed open. Who is correct?

    A. A only

    B. B only

    C. Both A and B

    D. Neither A nor B

35. Refer to the composite vehicle to answer this question: The ECM has a DTC for high crankcase pressure. Any of the following could be the cause EXCEPT:

    A. A damaged piston.

    B. Worn compression rings.

    C. A restricted crankcase ventilation tube.

    D. A restricted engine oil filter.

36. Refer to the composite vehicle (diagram on page 39 of the booklet) to answer this question: Which of the following would be the most likely cause of a 15-amp vehicle circuit breaker that has failed open?

    A. An open on circuit 178
    B. A short to ground on circuit 192
    C. An open on circuit 192
    D. A short to ground on circuit 178

37. Refer to the composite vehicle (diagram on page 37 of the booklet) to answer this question: The composite engine has two active DTCs, one for the intake manifold temperature sensor and one for the intake manifold pressure sensor. There are no other active or inactive codes. Technician A says an open on circuit 222 could be the cause. Technician B says an open at splice S25 could be the cause. Who is correct?

    A. A only
    B. B only
    C. Both A and B
    D. Neither A nor B

38. Refer to the composite vehicle to answer this question: Technician A says global maximum road speed is a programmable parameter. Technician B says engine idle speed is a programmable parameter. Who is correct?

    A. A only
    B. B only
    C. Both A and B
    D. Neither A nor B

39. Refer to the composite vehicle to answer this question: Technician A says a failed neutral safety switch may prevent a stationary regeneration of the DPF. Technician B says a failed accelerator can prevent a stationary regeneration of the DPF. Who is correct?

    A. A only
    B. B only
    C. Both A and B
    D. Neither A nor B

40. Refer to the composite vehicle to answer this question: All of the following are true concerning the air intake and exhaust system on the composite vehicle EXCEPT:

    A. The engine has a variable geometry turbocharger (VGT).
    B. The engine has a variable valve actuator (VVA).
    C. The engine has cooled exhaust gas recirculation (EGR).
    D. The engine has a DPF.

41. Refer to the composite vehicle to answer this question: The driver complains of poor performance from the air conditioner (A/C). The technician finds the engine cooling fan does not engage when the A/C is on. The fan will engage when commanded to from the scan tool. Which of the following is the most likely cause?

    A. A faulty engine cooling fan solenoid

    B. A failed engine cooling fan

    C. A failed blower motor switch

    D. A failed A/C high-side pressure switch

42. Refer to the composite vehicle to answer this question: The composite engine starts and runs but has low power. There are no active or inactive DTCs. Which of the following could be the cause?

    A. A missing signal from EPS1

    B. A missing signal from EPS2

    C. A broken injector return spring

    D. An open injector solenoid winding

43. Refer to the composite vehicle (diagram on page 39 of the booklet) to answer this question: The engine speed will increase appropriately when the remote power take-off (PTO) switch is operated; however, it will not operate when the dash-mounted PTO switch is operated. Which of the following could be the cause?

    A. An open circuit 170

    B. An open circuit 171

    C. An open at ECM pin 77

    D. An open at ECM pin 70

44. Refer to the composite vehicle to answer this question: The composite diesel engine has a DTC for high crankcase pressure. Which of the following could be the cause?

    A. A restricted crankcase ventilation filter

    B. Excessive engine oil pressure

    C. A missing crankcase ventilation filter

    D. An open on circuit 304

45. Refer to the composite vehicle (diagram on page 39 of the booklet) to answer this question: The fan will not operate from the fan control switch and the diagnostic switch will not cause the ECM to flash the DTCs. Which of the following could be the cause?

    A. A faulty fan control switch

    B. A faulty diagnostic switch

    C. An open at ECM pin 58

    D. An open circuit 161

Answer Keys and Explanations

INTRODUCTION

Included in this section are the answer keys for each preparation exam, followed by individual, detailed answer explanations and a reference identifying the designated task area being assessed by each specific question. This additional reference information may prove useful if you need to refer back to the task list located in Section 4 of this book for additional support.

PREPARATION EXAM 1 – ANSWER KEY

1.	A	21.	B	41.	D
2.	A	22.	C	42.	D
3.	D	23.	B	43.	D
4.	C	24.	D	44.	A
5.	D	25.	D	45.	C
6.	C	26.	B		
7.	C	27.	B		
8.	C	28.	D		
9.	C	29.	A		
10.	A	30.	B		
11.	B	31.	D		
12.	D	32.	D		
13.	C	33.	C		
14.	B	34.	C		
15.	A	35.	D		
16.	D	36.	B		
17.	D	37.	C		
18.	C	38.	C		
19.	B	39.	C		
20.	C	40.	B		

PREPARATION EXAM 1 – EXPLANATIONS

TASK A.1

1. A vehicle has been brought into the shop for diagnosis. Technician A says the engine serial number can be found on a data tag mounted on the engine. Technician B says the engine serial number is the last six digits of the vehicle identification number (VIN). Who is correct?

   A. A only
   B. B only
   C. Both A and B
   D. Neither A nor B

   Answer A is correct. Only Technician A is correct. The engine serial number is mounted on a tag on the engine. It is also usually stamped into the block.

   Answer B is incorrect. The VIN contains valuable information about the vehicle, but it does not have the engine serial number embedded in it.

   Answer C is incorrect. Only Technician A is correct.

   Answer D is incorrect. Technician A is correct.

TASK B.7

2. Technician A says the antilock brakes/electronic stability control systems (ABS/ESC) can cause the engine to derate. Technician B says the ABS/ESC system can cause the engine to shut off. Who is correct?

   A. A only
   B. B only
   C. Both A and B
   D. Neither A nor B

   Answer A is correct. Only Technician A is correct. If the ABS/ESC system senses wheel slippage under acceleration, it can command the engine to derate in an attempt to regain traction.

   Answer B is incorrect. The ABS/ESC system can derate the engine, but is not capable of shutting the engine off. The engine can be set to shut off in case of low coolant.

   Answer C is incorrect. Only Technician A is correct.

   Answer D is incorrect. Technician A is correct.

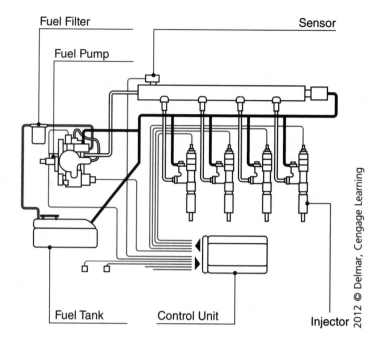

3. Referring to the figure above, which fuel system is illustrated?

TASK D.5

A.   Electronic unit injector (EUI)

B.   Hydraulically actuated electronic unit injector (HEUI)

C.   Pump line nozzle electronic (PLN-E)

D.   Common rail

Answer A is incorrect. An EUI fuel system will have rocker arms above the injectors, which provide the force to create the high pressure needed for injection.

Answer B is incorrect. An HEUI fuel system will have a high-pressure oil pump, which is used to create the high pressure needed for injection.

Answer C is incorrect. A PLN-E fuel system will have a mechanical injection pump, which creates the high pressure needed for injection.

Answer D is correct. This is a high-pressure common rail fuel system. The high pressure is created in the fuel pump. The injectors are energized by the electronic control module (ECM) to allow fuel flow into the combustion chamber.

TASK B.9

4. A connector lock is broken. Which of the following is the LEAST LIKELY acceptable repair method?

A. Replace only the lock.

B. Replace the lock and connector shell.

C. Use a nylon tie strap to secure the two halves of the connector together.

D. Replace the entire connector.

Answer A is incorrect. If the area where the lock engages the connectors is in good shape, replacing only the lock is an acceptable repair.

Answer B is incorrect. If the broken lock has damaged the injector shell, the connector shell may need to be replaced as well as the lock.

Answer C is correct. Using a nylon tie strap can possibly damage the wiring harness because of the pressure it will put on the wires as they lead into the connector. It is not considered an acceptable repair.

Answer D is incorrect. Some manufacturers do not provide individual pieces for a connector. In such a case, the only repair option may be to replace the entire connector with a complete replacement pigtail that includes wires, connector shell, and lock.

TASK C.2

5. Which of the following would be the LEAST LIKELY cause of high air intake temperature?

A. A restricted charge air cooler

B. A restricted radiator

C. A restricted condenser

D. A restricted air filter

Answer A is incorrect. A restricted charge air cooler can prevent sufficient cooling and cause high air intake temperature.

Answer B is incorrect. Because the charge air cooler and the radiator are often stacked one behind the other, a restricted radiator can reduce airflow across the charge air cooler and cause high air inlet temperatures.

Answer C is incorrect. Because the condenser is often stacked with the charge air cooler and the radiator, a restricted condenser can cause reduced airflow across the charge air cooler and result in high air intake temperatures,

Answer D is correct. A restricted air filter can cause low boost pressure, but will not cause high intake air temperatures.

6. Technician A says that the customer complaint should be verified before performing any work. Technician B says that verifying the complaint may include talking to the customer or driving the vehicle.

TASK A.2

   A. A only
   B. B only
   C. Both A and B
   D. Neither A nor B

Answer A is incorrect. Technician B is also correct.

Answer B is incorrect. Technician A is also correct.

Answer C is correct. Both Technicians are correct. It is important to verify the complaint. Perceived problems may be normal operational characteristics. Performance problems may require the vehicle to be driven.

Answer D is incorrect. Both Technicians are correct.

7. A high-mileage engine has low power and a misfire on cylinder #5. The technician finds an open injector coil on cylinder #5. After installing all new injectors, the engine has a misfire on cylinder #3. Which of the following could be the cause?

TASK B.10

   A. Low compression on cylinder #5
   B. Low compression on cylinder #3
   C. A damaged injector electrical pass-through connector on cylinder #3
   D. A damaged injector electrical pass-through connector on cylinder #5

Answer A is incorrect. Cylinder #5 was fixed with the new injector; there is no reason to believe that cylinder #5 has low compression.

Answer B is incorrect. Cylinder #3 did not have a misfire when the truck originally came in; there is no reason to believe that cylinder #3 now has low compression.

Answer C is correct. When replacing the injectors on high-mileage engines, it is common for the injector pass-through connectors to harden and then break when they are disconnected during injector replacement. The technician will most likely need to replace every injector pass-through connector to repair the engine properly.

Answer D is incorrect. A damaged injector pass-through connector on cylinder #5 would not cause cylinder #3 to have a misfire.

8. All of the following are styles of APP sensors EXCEPT:

TASK D.4

   A. A single potentiometer.
   B. A triple potentiometer.
   C. Optical.
   D. Hall effect.

Answer A is incorrect. Some manufacturers use single potentiometers as the APP. Some OEMs will also have an idle validation switch along with the single potentiometer.

Answer B is incorrect. Some manufacturers have two or three potentiometers as their APP.

Answer C is correct. No current production diesel engine manufacturers use an optical APP sensor.

Answer D is incorrect. Some manufacturers use a single or double Hall effect sensor APP.

TASK B.11

9. Which of the following would be the last step in a complete diagnosis and repair of a diesel engine?

A. Replace the faulty component.

B. Research service literature.

C. Verify the repair.

D. Verify the complaint.

Answer A is incorrect. Replacing the faulty component would occur near the end of the repair procedure. The technician would, however, need to verify whether the repair was effective after replacing the part.

Answer B is incorrect. Service literature should be reviewed early in the diagnostic process.

Answer C is correct. The last step in the repair of a diesel engine is to verify the repair. The technician needs to know that the original concern was repaired and no new concerns have arisen.

Answer D is incorrect. Verifying the complaint is the first step in repair, not the last.

TASK D.1

10. Which of the following is LEAST LIKELY to cause a no-start condition?

A. Incorrect idle speed programmed into the ECM

B. A failed engine position sensor

C. A failed ECM

D. Incorrect engine calibration files in the ECM

Answer A is correct. Incorrect idle speed programmed into the ECM will cause the engine to idle too fast, idle too slow, or possibly start and die. It will not cause a no-start.

Answer B is incorrect. A failed engine position sensor will not tell the ECM the position of the crankshaft; therefore, it can cause a no-start condition.

Answer C is incorrect. A failed ECM can prevent the injectors from being energized; this would cause a no-start condition.

Answer D is incorrect. Incorrect engine calibration files can cause the ECM to fail to recognize vital pieces of information coming from the sensors or cause the ECM to fail to command actuators, such as the injectors, to operate at the correct time. This could cause a no-start.

TASK B.3

11. When reprogramming an ECM, all of the following are important safety precautions EXCEPT:

A. The scan tool/laptop should be connected to a 110 volt AC (VAC) current.

B. Primary air pressure on the truck should be built to 120 psi.

C. The truck batteries should be fully charged.

D. The technician should verify that the latest flash file is being used.

Answer A is incorrect. The scan tool/laptop should be connected to 110 VAC. If the battery in the tool failed during the reprogramming, the ECM could be damaged.

Answer B is correct. Primary air pressure has no bearing on reprogramming an engine ECM and is not part of the safety precautions. The truck parking brake should be set, however.

Answer C is incorrect. The truck batteries should be fully charged. If the battery voltage falls too low during the process, the ECM could be damaged.

Answer D is incorrect. The technician should always install the latest reprogramming flash file when reprogramming an engine ECM.

12. Each of the following would be considered a mechanical engine problem EXCEPT:

    A. Low compression.

    B. Clogged diesel particulate filter (DPF).

    C. Leaking exhaust gas recirculation (EGR) cooler.

    D. Open injector wiring harness.

TASK A.3

Answer A is incorrect. Low compression can be caused by mechanical problems such as worn rings or leaking valves.

Answer B is incorrect. A clogged DPF must be cleaned. This is a mechanical condition. A failed sensor on the DPF would be an electrical condition.

Answer C is incorrect. A leaking EGR cooler will allow coolant into the air stream. This is a mechanical problem. The cooler will need to be replaced.

Answer D is correct. An open injector wiring harness is an electrical problem; it is preventing current flow to the injector. The wiring harness may be repaired or replaced as a unit, depending on the manufacturer's recommendation.

13. Technician A says a failure of the data bus could prevent the automated manual transmission from shifting properly. Technician B says a failure of the data bus could prevent the scan tool from communicating with the transmission controller. Who is correct?

    A. A only

    B. B only

    C. Both A and B

    D. Neither A nor B

TASK B.7

Answer A is incorrect. Technician B is also correct.

Answer B is incorrect. Technician A is also correct

Answer C is correct. Both Technicians are correct. A data bus failure can prevent the various modules from communicating with each other, thus preventing communication between the engine, transmission, and scan tool. The ABS will also fail to operate properly if the data bus has failed.

Answer D is incorrect. Both Technicians are correct.

14. Technician A says air inlet restriction should only be measured at idle. Technician B says a water manometer is used to measure air inlet restriction. Who is correct?

    A. A only

    B. B only

    C. Both A and B

    D. Neither A nor B

TASK C.1

Answer A is incorrect. Air inlet restriction should be measured when airflow is high while under full load.

Answer B is correct. Only Technician B is correct. A water manometer is the tool used to measure air inlet restriction. It can be the actual U-shaped tube or a gauge calibrated using the U-shaped tube.

Answer C is incorrect. Only Technician B is correct.

Answer D is incorrect. Technician B is correct.

TASK B.7

15. Technician A says a damaged throttle position sensor (TPS) can prevent proper EGR operation. Technician B says a damaged TPS can cause a no-start condition. Who is correct?

   A.   A only
   B.   B only
   C.   Both A and B
   D.   Neither A nor B

   Answer A is correct. Only Technician A is correct. The TPS communicates the position of the throttle valve to the ECM. The throttle valve produces a low-pressure area in the intake to enhance EGR flow. A damaged sensor would prevent proper operation of the valve and cause incorrect EGR flow.

   Answer B is incorrect. A failed TPS can affect power and EGR flow, but would not prevent the engine from starting.

   Answer C is incorrect. Only Technician A is correct.

   Answer D is incorrect. Technician A is correct.

TASK D.6

16. An engine equipped with an HEUI fuel system fails to start. A technician connects a scan tool to the engine. While cranking, the injection actuation pressure is 270 psi. Technician A says the fuel transfer pump is not providing sufficient pressure. Technician B says that the fuel filters should be checked for restriction. Who is correct?

   A.   A only
   B.   B only
   C.   Both A and B
   D.   Neither A nor B

   Answer A is incorrect. Injection actuation pressure is the oil pressure from the high-pressure oil pump. Injection actuation pressure of 270 psi is insufficient to allow the engine to start. Technician A should focus on the high-pressure oil system as the source of the no-start condition.

   Answer B is incorrect. Restricted fuel filters will cause low fuel pressure; however, the problem with this engine is that the injection actuation pressure is low.

   Answer C is incorrect. Neither Technician is correct.

   Answer D is correct. Neither Technician is correct.

TASK B.10

17. A diesel engine has several unrelated DTCs. During diagnosis, the technician measures AC voltage at the back of the alternator. Which of the following would be considered an acceptable voltage?

   A.   13.2 VAC
   B.   12.9 VAC
   C.   1.3 VAC
   D.   0.3 VAC

   Answer A is incorrect. A voltage reading of 13.2 VAC would be much more than the maximum specification of 0.5 VAC, indicating a failed alternator.

   Answer B is incorrect. A voltage reading of 12.9 VAC is an excessive amount of AC voltage from the alternator and can result in many DTCs being set.

   Answer C is incorrect. A voltage reading of 1.3 VAC is higher than specification and can cause the ECM to process various sensor readings incorrectly, resulting in various DTCs.

   Answer D is correct. A voltage reading of 0.3 VAC is lower than the specification of 0.5 VAC and would not normally be considered a problem.

2012 © Delmar, Cengage Learning

18. Referring to the figure above, the tool shown is used to:

    A. Measure diesel fuel-specific gravity.

    B. Measure coolant-specific gravity.

    C. Test the fuel system for air.

    D. Test the air intake system for diesel fuel.

    TASK D.2

    Answer A is incorrect. A refractometer is used to measure diesel fuel-specific gravity.

    Answer B is incorrect. A refractometer is used to measure coolant-specific gravity.

    Answer C is correct. This is a sight glass that is installed in the suction side of the fuel system prior to the transfer pump. It is used to check the fuel system for air leaks.

    Answer D is incorrect. Diesel fuel should not be in the air intake system. If it is, the diesel engine will suffer uncontrolled acceleration or hydraulic lock.

19. A vehicle has been brought to the shop multiple times to have the DPF cleaned using a stationary regeneration. Which of the following is the LEAST LIKELY cause?

    A. Incorrect ECM programming for the active regeneration of the after treatment device

    B. The use of ultralow-sulfur fuel

    C. The use of high-sulfur fuel

    D. Incorrect programming of the minimum vehicle speed parameter for active regeneration

    TASK B.11

    Answer A is incorrect. If the ECM is programmed not to allow active regeneration, this could be the cause.

    Answer B is correct. Engines with an after treatment system require the use of ultralow-sulfur diesel. This would not be the cause.

    Answer C is incorrect. High-sulfur fuel can cause DPF restriction and could be the cause.

    Answer D is incorrect. Minimum vehicle speed for active regeneration is normally set at about 30 mph. If the parameter was set at 75 mph, the truck probably would not be operated at that speed for a sufficient amount of time to allow active regeneration to occur.

20. Technician A says EUI injectors will need to be adjusted after installation. Technician B says that when replacing EUI injectors, the injector calibration code may need to be entered into the engine ECM. Who is correct?

    A. A only

    B. B only

    C. Both A and B

    D. Neither A nor B

    TASK D.7

    Answer A is incorrect. Technician B is also correct.

    Answer B is incorrect. Technician A is also correct.

    Answer C is correct. Both Technicians are correct. EUI injectors will need the installation height adjusted after installation. Some manufacturers require the use of a special height gauge. Others require the technician to bottom the plunger in the cup and then back the adjusting screw off a specified amount.

    Answer D is incorrect. Both Technicians are correct.

TASK B.1

21. Where would the after treatment diesel fuel injector be located?

    A. In the exhaust manifold before the turbocharger

    B. In the exhaust system after the turbocharger

    C. In the exhaust system between the diesel oxidation catalyst (DOC) and the DPF

    D. In the intake system prior to the EGR valve

Answer A is incorrect. The injector is not located before the turbocharger. This placement would result in raw fuel entering the turbo, which could damage it.

Answer B is correct. The injector is located after the turbocharger. This placement allows the fuel to raise the temperature in the DOC to start an active regeneration.

Answer C is incorrect. The injector is not located between the DOC and the DPF. If the injector's placement was after the DOC, it would not help raise the temperature of the DOC.

Answer D is incorrect. The injector is not in the intake system prior to the EGR valve. Spraying diesel fuel in the intake could result in a runaway situation.

TASK C.5

22. Technician A says glow plugs can be checked with an ohmmeter. Technician B says air inlet heaters can be checked with an amp clamp. Who is correct?

    A. A only

    B. B only

    C. Both A and B

    D. Neither A nor B

Answer A is incorrect. Technician B is also correct.

Answer B is incorrect. Technician A is also correct.

Answer C is correct. Both Technicians are correct. Glow plugs can be checked with an ohmmeter and will usually have a specification of less than 1 ohm. Air intake heaters can be checked with an amp clamp and will usually have a current draw in excess of 50 amps.

Answer D is incorrect. Both Technicians are correct.

TASK B.1

23. Each of the following is true concerning the exhaust after treatment system EXCEPT:

    A. It converts soot to ash.

    B. It may need to be removed from the truck to clean soot from the DPF.

    C. It needs to be removed from the truck to clean ash from the DPF.

    D. The ECM monitors the DPF for restriction with a differential pressure sensor.

Answer A is incorrect. The DPF converts soot to ash during a regeneration event.

Answer B is correct. Soot is cleaned from the filter by converting it to ash during the regeneration event. This happens with the filter on the truck.

Answer C is incorrect. Some DPF cleaners require the filter to be removed from the truck. Some cleaners also heat the filter in order to more effectively clean the filter.

Answer D is incorrect. The ECM monitors the pressure drop across the filter using a differential pressure sensor.

24. All of the following are methods of DPF regeneration EXCEPT:

    A.   Passive.

    B.   Active.

    C.   Stationary.

    D.   Engine off.

TASK C.6

Answer A is incorrect. Passive regeneration occurs when the exhaust system is hot enough to clean soot from the DPF during normal engine operation.

Answer B is incorrect. Active regeneration occurs when the ECM determines that the DPF is restricted and needs to be cleaned, the vehicle speed is above 25 mph, and the exhaust temperature is not hot enough for regeneration. The ECM will inject fuel into the exhaust stream to raise exhaust temperature enough to allow regeneration to occur.

Answer C is incorrect. Stationary regeneration occurs when the vehicle is parked and is commanded to enter regeneration either with a dash-mounted switch or a scan tool.

Answer D is correct. DPF regeneration cannot occur when the engine is off.

25. A diesel engine has had the second set of injectors replaced in less than 10,000 miles. Both sets of injectors had excessive internal leakage. Technician A says the problem could be a restricted air filter. Technician B says the problem could be a restricted DPF. Who is correct?

    A.   A only

    B.   B only

    C.   Both A and B

    D.   Neither A nor B

TASK D.8

Answer A is incorrect. A restricted air filter can cause low power, but would not cause the injectors to wear prematurely.

Answer B is incorrect. A restricted DPF can cause low power, but will not cause the injectors to wear prematurely.

Answer C is incorrect. Neither Technician is correct.

Answer D is correct. Neither Technician is correct. When injectors wear prematurely in a diesel engine, the technician should inspect the fuel filtration system and the fuel handling techniques employed by the customer.

26. Refer to the composite vehicle to answer this question: The composite diesel engine blows black smoke under acceleration. Which of the following is the LEAST LIKELY cause?

    A.   A damaged DPF

    B.   A restricted exhaust

    C.   Worn injectors

    D.   Incorrect injector calibration codes

TASK A.4

Answer A is incorrect. The DPF is supposed to trap particulate emissions and eliminate exhaust smoke. If the exhaust is black under acceleration, then the DPF is failing.

Answer B is correct. A restricted exhaust can cause low power, but should not cause black smoke.

Answer C is incorrect. Worn injectors can give poor spray patterns and result in overfueling, a damaged DPF, and black exhaust smoke.

Answer D is incorrect. Incorrect injector calibration codes installed in the ECM can cause overfueling, a damaged DPF, and black exhaust smoke.

TASK B.1

27. Refer to the composite vehicle to answer this question: The technician has retrieved a diagnostic trouble code (DTC) for a failed DPF. When checking the voltage with the engine running under various loads and RPMs, the after treatment DPF differential pressure sensor signal voltage never changes from 0.8 volts. Which of the following could be the cause?

   A. A restricted DPF
   B. A missing DPF
   C. A restricted EGR passage
   D. A stuck EGR valve

   Answer A is incorrect. A restricted DPF will cause the voltage signal to be high, and the signal will vary under engine load.

   Answer B is correct. If the DPF has been removed, the pressure between the two sensing ports of the DPF will not change. Thus, the voltage will not change.

   Answer C is incorrect. A restricted EGR passage will result in low EGR flow, but will not cause the after treatment diesel particulate filter differential pressure sensor signal voltage to remain constant.

   Answer D is incorrect. A stuck EGR valve can cause insufficient or excessive EGR flow, but will not cause after treatment DPF differential pressure sensor signal voltage to remain low and constant.

TASK B.2

28. Refer to the composite vehicle (diagram on page 40 of the booklet) to answer this question: The ECM will not communicate with the transmission module. Which data bus could be the cause?

   A. J1708
   B. J1587
   C. J1922
   D. J1939

   Answer A is incorrect. J1708 is the communications protocol. It is not a data bus.

   Answer B is incorrect. The ECM will communicate with the scan tool on J1587, but does not use this data bus to communicate with other modules.

   Answer C is incorrect. J1922 is **not** used on the composite engine.

   Answer D is correct. The other modules on the truck communicate with the engine ECM over J1939. The scan tool can also communicate to the ECM over J1939.

TASK B.3

29. Refer to the composite vehicle to answer this question: Technician A says the idle shutdown timer is an adjustable parameter on the composite engine. Technician B says that maximum engine RPM is an adjustable parameter on the composite engine. Who is correct?

   A. A only
   B. B only
   C. Both A and B
   D. Neither A nor B

   Answer A is correct. Only Technician A is correct. The idle shutdown timer can be adjusted in the range of 1 to 100 minutes.

   Answer B is incorrect. Maximum engine RPM is part of the engine calibration file. It may be changed if the engine is recalibrated, but it is not a customer-adjustable parameter.

   Answer C is incorrect. Only Technician A is correct.

   Answer D is incorrect. Technician A is correct.

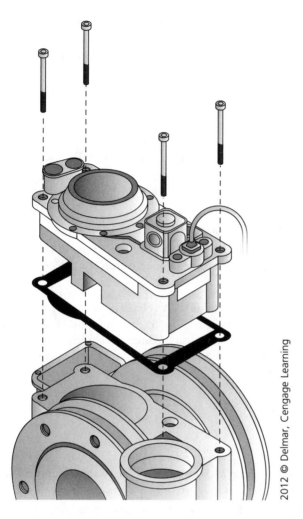

2012 © Delmar, Cengage Learning

30. Refer to the composite vehicle for this question: Referring to the figure above, what component is being removed?

TASK C.3

   A.   The after treatment injector

   B.   The variable geometry turbocharger (VGT) actuator

   C.   The boost pressure sensor

   D.   The diesel particulate filter (DPF)

Answer A is incorrect. The after treatment injector is located after the turbocharger. This is the VGT actuator.

Answer B is correct. This is the VGT actuator. It controls the VGT turbocharger.

Answer C is incorrect. This is not the boost pressure sensor. The boost pressure sensor would be located in or around the intake manifold.

Answer D is incorrect. This is not the DPF. The DPF is located in the exhaust system.

TASK B.4

31. Refer to the composite vehicle to answer this question: The EGR valve fails to operate. There is a signal voltage to the valve at the IX connector. Technician A says the EGR cooler may be restricted. Technician B says the EGR venturi may be restricted. Who is correct?

   A. A only
   B. B only
   C. Both A and B
   D. Neither A nor B

Answer A is incorrect. A restricted EGR cooler can cause insufficient EGR flow, but would not cause the valve to fail to operate.

Answer B is incorrect. The EGR venturi can cause low EGR flow, but will not cause the EGR valve to be inoperative.

Answer C is incorrect. Neither Technician is correct.

Answer D is correct. Neither Technician is correct. The most likely problem is that the EGR valve has failed mechanically. It may need to be cleaned or replaced, depending on the manufacturer's recommendations.

TASK B.5

32. Refer to the composite vehicle to answer this question: A vehicle has been sitting in the shop for more than 24 hours. A technician connects a scan tool to the engine and finds the coolant temperature and air inlet temperature to have identical readings. Technician A says the coolant temperature sensor is faulty and should be replaced. Technician B says a failed ECM can cause this condition. Who is correct?

   A. A only
   B. B only
   C. Both A and B
   D. Neither A nor B

Answer A is incorrect. After a vehicle sits for 24 hours, the coolant temperature and air inlet temperature should have the same reading. If they are different, it could indicate a problem.

Answer B is incorrect. A failed ECM could cause a temperature signal to be incorrectly interpreted and incorrectly displayed; however, in this case it is a normal condition for the coolant and air inlet temperatures to have identical readings; no failure is indicated.

Answer C is incorrect. Neither Technician is correct.

Answer D is correct. Neither Technician is correct. The readings indicate a normal condition; using a scan tool is a valid test method.

TASK B.6

33. Refer to the composite vehicle (diagram on page 39 of the booklet) to answer this question: Technician A says the ECM receives unswitched battery power on Connector 3 ECM Pin 79. Technician B says the ECM receives switched battery power on Connector 3 ECM Pin 78. Who is correct?

   A. A only
   B. B only
   C. Both A and B
   D. Neither A nor B

Answer A is incorrect. Technician B is also correct.

Answer B is incorrect. Technician A is also correct.

Answer C is correct. Both Technicians are correct. Pin 79 is unswitched battery power through a 15 amp circuit breaker. Pin 78 is switched battery power through the normally open power relay.

Answer D is incorrect. Both Technicians are correct.

34. Refer to the composite vehicle to answer this question: All of the following are acceptable voltage drop measurements on the ECM unswitched battery power wire Circuit 179 EXCEPT:

    A. 0.1 VDC.
    B. 0.01 VDC.
    C. 1.5 VDC.
    D. 0.15 VDC.

    TASK B.8

    Answer A is incorrect. A small voltage drop of 0.1 VDC is acceptable.

    Answer B is incorrect. A voltage drop of 0.01 VDC is extremely small and indicates a very good circuit.

    Answer C is correct. A voltage drop of 1.5 VDC is excessive and could cause the ECM to have insufficient voltage to operate the injectors, resulting in a misfire.

    Answer D is incorrect. A voltage drop of 0.15 VDC is well below the normal standard of 0.5 VDC maximum voltage drop.

35. Refer to the composite vehicle for this question: Under what conditions is the high exhaust temperature sensor (HETS) lamp illuminated?

    A. The diesel particulate filter regeneration (DPFR) status lamp is illuminated.
    B. Exhaust gas temperature 1 (EGT1) is above 850°F (454.4°C) and vehicle speed is below 5 mph.
    C. EGT2 is above 850°F and vehicle speed is below 5 mph.
    D. EGT3 is above 850°F and vehicle speed is below 5 mph.

    TASK C.4

    Answer A is incorrect. The DPFR lamp is illuminated when the DPF needs to be regenerated. The ECM calculates DPF restriction based on pressure sensor differentials.

    Answer B is incorrect. EGT1 is not used to illuminate the HETS lamp.

    Answer C is incorrect. EGT2 is not used to illuminate the HETS lamp.

    Answer D is correct. The ECM will illuminate the HETS lamp when EGT3 is above 850°F and vehicle speed is below 5 mph. EGT3 is used because it is located closest to the exhaust outlet. The HETS lamp is a safety feature to notify the driver of elevated exhaust temperatures at low speeds or when the vehicle is stationary.

36. Refer to the composite vehicle to answer this question: The engine ECM has a DTC for an open coolant temperature sensor. The technician disconnects the wiring harness at connector F and jumpers across the two terminals. The ECM now sets a code for a shorted coolant temperature sensor. Which of the following could be the cause?

    TASK B.2

    A. A faulty ECM
    B. A faulty coolant temperature sensor
    C. A faulty wiring harness
    D. A faulty air intake temperature sensor

    Answer A is incorrect. The ECM is not faulty. The tests proved that the ECM can diagnose open and shorted circuits.

    Answer B is correct. The coolant temperature sensor is faulty. It is electrically open. When the technician jumpered the two terminals, the ECM set a shorted code that verified that the wiring harness and ECM are OK. Therefore, the only possible cause of the trouble code is a faulty coolant temperature sensor.

    Answer C is incorrect. If the wiring harness was faulty, it would not have set a shorted code when the technician jumpered the two terminals.

    Answer D is incorrect. The air intake temperature sensor was not the circuit being tested.

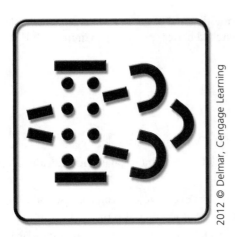

2012 © Delmar, Cengage Learning

TASK B.6

37. Refer to the composite vehicle to answer this question: Referring to the figure above, the DPF status lamp is flashing on the dash of the composite vehicle. Which of the following is indicated?

   A.   The air intake filter is restricted.

   B.   The cabin air filter is restricted.

   C.   DPF filter regeneration is necessary because there is a moderate soot load.

   D.   The truck should be parked immediately and towed to a shop.

   Answer A is incorrect. This lamp is not an air intake restriction lamp. The air intake restriction gauge should be checked to know the status of the air filter.

   Answer B is incorrect. The cabin air filter does not typically have a dash lamp.

   Answer C is correct. The DPF needs regeneration. The truck can be driven at sustained highway speeds, or it can be parked and a stationary regeneration performed.

   Answer D is incorrect. The truck should be parked, however, if this lamp is on along with both the check engine and stop engine lamps.

TASK B.2

38. Refer to the composite vehicle to answer this question: At what temperature will the ECM derate the engine by 60 percent?

   A.   225°F (107.2°C)

   B.   230°F (110°C)

   C.   235°F (112.8°C)

   D.   240°F (115.6°C)

   Answer A is incorrect. The ECM will derate the engine 20 percent at 225°F (107.2°C).

   Answer B is incorrect. The ECM will derate the engine 40 percent at 230°F (110°C).

   Answer C is correct. The ECM will derate the engine 60 percent at 235°F (112.8°C).

   Answer D is incorrect. The ECM will shut down the engine at 240°F (115.6°C).

39. Refer to the composite vehicle to answer this question: Which of the following tools is LEAST LIKELY to be used when checking the data bus?

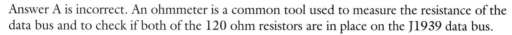

A. Ohmmeter

B. Voltmeter

C. Ammeter

D. Scan tool

TASK B.7

Answer A is incorrect. An ohmmeter is a common tool used to measure the resistance of the data bus and to check if both of the 120 ohm resistors are in place on the J1939 data bus.

Answer B is incorrect. A voltmeter can be used to measure high-side and low-side voltages.

Answer C is correct. There are no specifications available to perform a current flow test on the data bus.

Answer D is incorrect. A scan tool can be used to check if the various modules can communicate via the data bus.

40. Refer to the composite vehicle to answer this question: Cylinder #5 has a misfire. During diagnosis, the technician measures the resistance of the injector solenoid. Which of the following resistance readings indicates a good injector solenoid?

A. 0.4 ohms

B. 4.0 ohms

C. 0.2 ohms

D. 22.0 ohms

TASK B.8

Answer A is incorrect. The minimum resistance is 0.5 ohms. A resistance reading of 0.4 ohms would indicate a failed solenoid.

Answer B is correct. The resistance specification is 0.5 to 5.0 ohms. Therefore, 4.0 ohms would indicate a good solenoid.

Answer C is incorrect. A resistance reading of 0.2 ohms is less than the minimum specification and would indicate a failed solenoid.

Answer D is incorrect. A resistance reading of 22 ohms is higher than the maximum specification of 5.0 ohms and would indicate a failed solenoid.

41. Refer to the composite vehicle to answer this question: The speedometer is displaying a faster vehicle speed than the vehicle is actually traveling. All of the following could cause this EXCEPT:

A. An incorrect rear axle ratio setting in the ECM.

B. An incorrect tire size rev/mile setting in the ECM.

C. An incorrect transmission tail shaft gear setting in the ECM.

D. An incorrect maximum cruise control setting in the ECM.

TASK B.7

Answer A is incorrect. If the rear axle ratio setting was set incorrectly, the speedometer would incorrectly calculate vehicle speed.

Answer B is incorrect. If the tire size was programmed incorrectly, the ECM would have incorrect information and calculate vehicle speed incorrectly.

Answer C is incorrect. If the number of teeth on the tail shaft was set higher than the actual number of teeth, the vehicle speed would display faster than actual speed.

Answer D is correct. The cruise control settings will affect the cruise control operation, but will not affect the speedometer display.

TASK B.3

42. Refer to the composite vehicle for this question: Technician A says the idle engine speed programmable parameter can be set between 600 and 950 rpm. Technician B says the maximum power take-off (PTO) speed can be set between 500 and 850 rpm. Who is correct?

    A. A only
    B. B only
    C. Both A and B
    D. Neither A nor B

    Answer A is incorrect. The idle speed parameter range is between 600 and 850 rpm.

    Answer B is incorrect. The maximum PTO speed is adjustable between 600 and 2500 rpm.

    Answer C is incorrect. Neither Technician is correct.

    Answer D is correct. Neither Technician is correct.

TASK D.3

43. Refer to the composite vehicle to answer this question: Which of the following best describes the fuel system on the composite vehicle?

    A. Common rail
    B. HEUIs
    C. PLN-E
    D. EUIs

    Answer A is incorrect. In a common rail system, the common rail is filled with fuel pressurized to approximately 20,000 psi. This is an EUI system.

    Answer B is incorrect. An HEUI system will have a high-pressure oil pump for the injectors. This is an EUI system.

    Answer C is incorrect. A PLN-E system will have a mechanical injection pump. This is an EUI system.

    Answer D is correct. This is an EUI system; it has camshaft-operated EUIs.

TASK B.5

44. Refer to the composite vehicle to answer this question: Technician A says the engine oil pressure sensor provides the information for the dash gauge. Technician B says the accelerator pedal position (APP) sensor has three potentiometers. Who is correct?

    A. A only
    B. B only
    C. Both A and B
    D. Neither A nor B

    Answer A is correct. Only Technician A is correct. The engine oil pressure sensor provides information for both the dash-mounted gauge and the engine ECM.

    Answer B is incorrect. The APP sensor has two potentiometers, not three.

    Answer C is incorrect. Only Technician A is correct.

    Answer D is incorrect. Technician A is correct.

45. Refer to the composite vehicle for this question: The stop engine light on the composite engine started flashing and 30 seconds later the engine shut down. Which of the following could be the cause?

TASK B.6

   A. A failed APP sensor
   B. A failed after treatment DPF inlet temperature sensor (EGT2)
   C. High engine oil temperature
   D. Low engine coolant temperature

Answer A is incorrect. A failed APP sensor may prevent the engine from accelerating, but it will not cause the ECM to enter into engine protection mode.

Answer B is incorrect. A failed after treatment DPF inlet temperature sensor (EGT2) can prevent DPF regeneration, but will not cause the engine to enter into protection mode.

Answer C is correct. High engine oil temperature or high coolant temperature can cause the engine to enter into protection mode.

Answer D is incorrect. Low coolant temperature will not cause the engine to enter into protection mode, but it can cause the engine to idle at a higher than normal RPM.

PREPARATION EXAM 2 – ANSWER KEY

1.	B	**21.**	D	**41.**	A
2.	D	**22.**	D	**42.**	D
3.	D	**23.**	B	**43.**	D
4.	A	**24.**	D	**44.**	C
5.	A	**25.**	D	**45.**	B
6.	C	**26.**	B		
7.	B	**27.**	C		
8.	B	**28.**	C		
9.	C	**29.**	D		
10.	C	**30.**	C		
11.	A	**31.**	A		
12.	C	**32.**	D		
13.	B	**33.**	D		
14.	B	**34.**	D		
15.	A	**35.**	A		
16.	C	**36.**	A		
17.	B	**37.**	C		
18.	B	**38.**	B		
19.	A	**39.**	C		
20.	C	**40.**	C		

PREPARATION EXAM 2 – EXPLANATIONS

TASK A.8

1. All of the following can cause an immediate reduction in boost pressure EXCEPT:

   A. Leaking charge air cooler hoses.

   B. A leaking air filter gasket.

   C. A restricted air filter.

   D. A restricted diesel particulate filter (DPF).

   Answer A is incorrect. Leaking charge air cooler hoses will allow the compressed air to escape and result in low boost pressure.

   Answer B is correct. A leaking air filter gasket will allow dirty air to enter the engine, but since it is before the turbocharger, it will not cause an immediate reduction in boost pressure. Over time, however, the dirty air will wear the turbocharger compressor wheel and can result in reduced boost pressure.

   Answer C is incorrect. A restricted air filter impedes airflow and will lower boost pressure.

   Answer D is incorrect. A restricted DPF impedes airflow and will cause reduced boost pressure.

2. A truck with an inactive DTC for EGT1 out of range is being repaired. After sitting overnight, a scan tool is connected to the engine. EGT1 indicates 80°F (26.7°C); EGT2 indicates 70°F (21.1°C); EGT3 indicates 75°F (23.9°C). Which of the following is the most likely cause of the code?

    TASK B.2

    A. EGT1 is faulty.

    B. EGT2 is faulty.

    C. EGT3 is faulty.

    D. A high-resistance electrical connection on EGT1.

    Answer A is incorrect. Since EGT1, 2, and 3 are all indicating the same approximate temperature, they are probably working correctly at this time.

    Answer B is incorrect. All the exhaust gas temperature sensors are indicating practically the same temperature; they appear to be functioning normally at this time.

    Answer C is incorrect. Since all three sensors are indicating nearly the same temperature and the code is inactive, the temperature sensors probably are indicating the correct temperatures.

    Answer D is correct. The technician should monitor the sensor signals while moving the wires in the EGT1 circuit. The problem could be a loose wire. If no problem is found, the technician should measure the resistance at the sensor and the wiring harness.

3. Which of the following is LEAST LIKELY to cause a low power concern?

    TASK B.7

    A. A faulty ABS controller

    B. A restricted DPF

    C. A seized engine cooling fan clutch

    D. A faulty dash-mounted ECM diagnostic switch

    Answer A is incorrect. A faulty ABS controller can send a signal to the ECM to reduce torque.

    Answer B is incorrect. A restricted DPF can increase backpressure and lower engine power.

    Answer C is incorrect. A seized fan clutch will cause the fan to turn all the time; this will lower engine power to the wheels and increase fuel consumption.

    Answer D is correct. A faulty dash-mounted diagnostic switch can prevent the operator from retrieving DTCs, but will not affect engine power output.

4. Which of the following is LEAST LIKELY to be associated with a stuck open EGR valve?

    TASK C.7

    A. Excessive boost pressure

    B. Poor acceleration

    C. Excessive fuel pressure

    D. Low fuel pressure

    Answer A is correct. A stuck open EGR valve will not cause excessive boost pressure, though it can cause low power.

    Answer B is incorrect. A stuck open EGR valve can result in poor acceleration because it will allow excessive exhaust gas to flow into the intake.

    Answer C is incorrect. The EGR valve does not affect fuel pressure; it affects incoming airflow.

    Answer D is incorrect. The EGR valve will not affect fuel pressure; however, low fuel pressure can cause low power.

TASK B.3

5. A diesel engine dies after idling for three minutes. The engine will not restart. Technician A says low fuel level could be the cause. Technician B says the idle shutdown timer in the electronic control module (ECM) could be the cause. Who is correct?

   A. A only
   B. B only
   C. Both A and B
   D. Neither A nor B

   Answer A is correct. Only Technician A is correct. If the engine does not restart, the problem could be that it is out of fuel.

   Answer B is incorrect. If the idle shutdown timer causes the engine to die, it should restart.

   Answer C is incorrect. Only Technician A is correct.

   Answer D is incorrect. Technician A is correct.

TASK B.2

6. Technician A says an ECM image can be electronically stored with the work order. Technician B says an ECM image can be saved to a file. Who is correct?

   A. A only
   B. B only
   C. Both A and B
   D. Neither A nor B

   Answer A is incorrect. Technician B is also correct.

   Answer B is incorrect. Technician A is also correct.

   Answer C is correct. Both Technicians are correct. The image can be saved to a work order to verify repairs in case of a future concern with the vehicle; the image can also be saved to a file for later reference in a repair.

   Answer D is incorrect. Both Technicians are correct.

TASK C.9

7. A DTC has been set for the variable valve actuator on a late-model diesel engine. Technician A says the variable valve actuator opens the exhaust valve. Technician B says the variable valve actuator opens the intake valve. Who is correct?

   A. A only
   B. B only
   C. Both A and B
   D. Neither A nor B

   Answer A is incorrect. The variable valve actuator opens the intake valve, not the exhaust valve.

   Answer B is correct. Only Technician B is correct. The variable valve actuator opens the intake valve to help improve airflow and lower emissions.

   Answer C is incorrect. Only Technician B is correct.

   Answer D is incorrect. Technician B is correct.

After treatment diesel oxidation catalyst (DOC) inlet temperature	700°F (371.1°C)
After treatment diesel particulate filter (DPF) inlet temperature	600°F (315.6°C)
After treatment diesel particulate filter (DPF) outlet temperature	500°F (260°C)

8. The readings shown above were taken 10 minutes after a stationary regeneration was started. Which of the following is indicated?

TASK B.5

A. A restricted DPF

B. A damaged DOC

C. A damaged DPF

D. A damaged DPF outlet temperature sensor

Answer A is incorrect. The stationary regeneration is performed to clean a restricted DPF. These temperatures are lower than normal. The after treatment system is not warming the DPF sufficiently.

Answer B is correct. A damaged DOC could be indicated. The DOC should raise the temperature of the air entering the DPF.

Answer C is incorrect. The temperatures are lower than expected; it is the job of the DOC to raise the temperature of the air going into the DPF. The DOC, not the DPF, is most likely the problem.

Answer D is incorrect. Since both DPF sensors show lower than normal temperatures, it is unlikely that the DPF outlet sensor is faulty.

9. All of the following could cause a diesel engine equipped with an electronic unit injector (EUI) fuel system to crank but not start EXCEPT:

TASK B.4

A. Low fuel level.

B. A failed ECM.

C. Low oil pressure.

D. A failed engine position sensor.

Answer A is incorrect. Low fuel level can cause a crank but no-start condition. The technician should visually check the level in the tank rather than relying on the dash gauges.

Answer B is incorrect. A failed ECM can fail to turn the injectors on. This would cause a crank but no-start concern.

Answer C is correct. Low oil pressure can cause a no-start on a hydraulic electronic unit injector (HEUI) engine because oil pressure is needed to operate the injectors. It could only cause a start and shut-off complaint on an EUI engine, however, because an EUI engine does not require oil pressure to operate the injectors. The injectors on an EUI engine are camshaft operated.

Answer D is incorrect. If the ECM does not receive an engine position sensor signal, the engine can crank but not start.

TASK B.1

10. During a stationary regeneration of the DPF, all of the following would normally occur EXCEPT:

A. Turbo boost would increase.

B. Exhaust temperature will rise.

C. A DTC will be set for the DPF.

D. Fuel will be injected into the exhaust stream.

Answer A is incorrect. The ECM will increase the boost pressure by operating the variable geometry turbocharger (VGT).

Answer B is incorrect. The exhaust temperature will rise to approximately 900°F (482.2°C) in order to convert the soot to ash.

Answer C is correct. If a DTC is set, it may indicate a missing or damaged DPF.

Answer D is incorrect. Fuel will be injected in the exhaust stream to raise the temperature inside the DPF.

TASK D.2

11. Technician A says it may be necessary to loosen a bleeder screw when removing air from the fuel system. Technician B says it is standard procedure to crank the engine for four minutes at a time when bleeding the air from the fuel system. Who is correct?

A. A only

B. B only

C. Both A and B

D. Neither A nor B

Answer A is correct. Only Technician A is correct. Since air is compressible, it may be necessary to loosen a bleeder screw or fitting to allow the air to escape.

Answer B is incorrect. The normal cranking recommendations are 30 seconds at a time, with two minutes in between for the starter to cool down.

Answer C is incorrect. Only Technician A is correct.

Answer D is incorrect. Technician A is correct.

TASK B.2

12. Which of the following problems is LEAST LIKELY to set a DTC?

A. An open after treatment injector solenoid

B. A shorted after treatment injector solenoid

C. An open engine brake on/off switch

D. A shorted engine position sensor 2 (EPS2)

Answer A is incorrect. An open after treatment injector solenoid would not operate the injector when commanded by the ECM; the ECM would see no current flow and set a DTC.

Answer B is incorrect. A shorted after treatment injector solenoid would cause a high current flow and the ECM would set a DTC.

Answer C is correct. The engine brake on/off switch has two positions, open and closed; therefore, a switch that failed open would not be an abnormal operating condition and would not set a DTC.

Answer D is incorrect. A shorted EPS2 would not send a signal. The ECM would compare EPS1 to EPS2 and realize that EPS2 was malfunctioning and set a DTC.

13. A vehicle has multiple counts of an inactive DTC for "coolant temperature sensor signal erratic." Which of the following is the most likely cause?

TASK B.9

   A. Loose connector pins on the intake manifold temperature sensor
   B. Loose connector pins on the coolant temperature sensor
   C. A faulty intake manifold temperature sensor
   D. A faulty coolant temperature sensor

Answer A is incorrect. Loose connector pins will set an erratic DTC; however, the DTC set is for the coolant temperature sensor, not the intake manifold temperature sensor.

Answer B is correct. Loose connector pins on the coolant temperature sensor could cause momentary changes in the signal, resulting in an erratic DTC being set for the circuit.

Answer C is incorrect. The DTC is set for the coolant temperature sensor, not the intake manifold temperature sensor.

Answer D is incorrect. The coolant temperature sensor may be faulty, but normally this type of DTC is set by loose connections in the circuit.

2012 © Delmar, Cengage Learning

14. Referring to the figure above, the lamp shown is illuminated on the dash. Which of the following is indicated?

TASK B.6

   A. High exhaust restriction
   B. High exhaust temperature
   C. High intake restriction
   D. High intake temperature

Answer A is incorrect. This is the temperature lamp, not the restriction lamp.

Answer B is correct. This lamp alerts the driver to a high exhaust temperature condition. A regeneration is in progress; it does not indicate the need for immediate service.

Answer C is incorrect. This indicates high exhaust temperature, not high intake restriction.

Answer D is incorrect. This indicates high exhaust temperature, not high intake temperature.

TASK B.7

15. The engine in a vehicle will not start. Technician A says an anti-theft security device could prevent the engine from starting. Technician B says the antilock brake system (ABS) controller could prevent the engine from starting. Who is correct?

   A. A only

   B. B only

   C. Both A and B

   D. Neither A nor B

Answer A is correct. Only Technician A is correct. A vehicle's security system can prevent the engine from starting if the incorrect code is entered into the on-dash input device.

Answer B is incorrect. The ABS controller can reduce power, but cannot prevent the engine from starting.

Answer C is incorrect. Only Technician A is correct.

Answer D is incorrect. Technician A is correct.

TASK B.5

16. Technician A says most ECMs will save freeze frame data on active codes. Technician B says most ECMs will save freeze frame data on inactive codes. Who is correct?

   A. A only

   B. B only

   C. Both A and B

   D. Neither A nor B

Answer A is incorrect. Technician B is also correct.

Answer B is incorrect. Technician A is also correct.

Answer C is correct. Both Technicians are correct. Freeze frame data is stored when a code becomes active, and will continue to be saved if the code becomes inactive.

Answer D is incorrect. Both Technicians are correct.

TASK B.11

17. The technician has replaced a failed head gasket. After the test drive to confirm that the repair was successful, the ECM has an active trouble code for a damaged DPF. Which of the following is the most likely cause?

   A. A failed injector

   B. Damage to the DPF due to coolant from the failed head gasket

   C. A leaking charge air cooler

   D. A failed EGR valve

Answer A is incorrect. While a failed injector can cause a failed DPF, there is no evidence that the injector has failed.

Answer B is correct. If the failed head gasket has allowed substantial coolant flow to the DPF for an extended period of time, it is possible that the DPF has been damaged.

Answer C is incorrect. A leaking charge air cooler will result in low power, but should not damage the DPF.

Answer D is incorrect. A failed EGR valve can cause low power, but should not damage the DPF.

18. The remote power take-off (PTO) switch will not engage the remote PTO function. A technician has connected a scan tool to the engine and finds that the ECM indicates remote PTO "OFF" regardless of switch position. All of the following could be a cause EXCEPT:

TASK B.5

  A. No voltage supplied to the PTO switch.

  B. A failed PTO drive gear.

  C. A failed PTO switch.

  D. A failed ECM.

Answer A is incorrect. No voltage to the remote PTO switch would mean no signal would be sent to the ECM.

Answer B is correct. A failed drive gear may prevent the PTO from operating, but it would not prevent the ECM from receiving the signal from the switch.

Answer C is incorrect. A failed PTO switch could possibly not send a signal to the ECM.

Answer D is incorrect. A failed ECM may not recognize a signal from the remote PTO switch.

19. The scan tool will not communicate with the engine. Which of the following is LEAST LIKELY cause?

TASK B.4

  A. Low fuel level in the fuel tank

  B. No ground at the ECM

  C. No battery positive at the ECM

  D. A failed data bus

Answer A is correct. Low fuel level can cause a no-start condition. However, it will not prevent the ECM from communicating with the scan tool.

Answer B is incorrect. No ground at the ECM can prevent the ECM from operating, resulting in no communications.

Answer C is incorrect. No battery positive power at the ECM can prevent the ECM from communicating. This can be caused by a blown fuse.

Answer D is incorrect. A failed data bus can prevent the ECM from communicating. It is still possible that the engine may start.

20. Technician A says the EGR valve helps reduce NOx emissions. Technician B says the DPF helps reduce particulate emissions. Who is correct?

TASK B.1

  A. A only

  B. B only

  C. Both A and B

  D. Neither A nor B

Answer A is incorrect. Technician B is also correct.

Answer B is incorrect. Technician A is also correct.

Answer C is correct. Both Technicians are correct. The EGR valve and the diesel exhaust fluid system both help lower NOx emissions. The DPF traps particulates and converts them to ash.

Answer D is incorrect. Both Technicians are correct.

TASK B.4

21. A diesel engine equipped with an EUI fuel system has a misfire on cylinder #2. The technician replaces the injector and the cylinder continues to misfire. Which of the following is the LEAST LIKELY cause?

  A. A worn cam lobe
  B. A failed injector harness
  C. A burned exhaust valve
  D. Low fuel supply pressure

Answer A is incorrect. A worn cam lobe could be the cause. EUI fuel systems use the cam lobe to create the pressure for injection.

Answer B is incorrect. Failed injector harnesses are relatively common. If the harness was not supplying voltage to the injector, that could cause the misfire.

Answer C is incorrect. If the exhaust valve was burned, the cylinder would be low on compression, causing a misfire.

Answer D is correct. If the fuel supply pressure was low it would affect all cylinders, not just a cylinder in the middle of the engine.

TASKS A.7, B.2

22. The composite diesel engine has an inactive diagnostic trouble code (DTC) for high engine oil temperature. The freeze frame data indicates the engine coolant temperature was 160°F (71°C) and the intake manifold temperature was 100°F (37.8°C) when the condition occurred. Which of the following is the most likely cause of the stored DTC?

  A. A restricted radiator
  B. A restricted exhaust gas recirculation (EGR) cooler
  C. A faulty coolant temperature sensor
  D. A faulty oil temperature sensor

Answer A is incorrect. A restricted radiator would have caused the coolant temperature to be higher.

Answer B is incorrect. A restricted EGR cooler would cause EGR temperature to be higher than normal; it would not cause engine oil temperature to be higher than normal.

Answer C is incorrect. A faulty coolant temperature sensor could cause a coolant temperature DTC, but not an oil temperature DTC.

Answer D is correct. Since the coolant temperature and intake manifold temperatures were low, the most likely cause is a faulty oil temperature sensor or wiring harness.

TASK B.3

23. Refer to the composite vehicle to answer this question: Which of the following is the minimum speed at which a passive regeneration may occur?

  A. 10 mph
  B. 25 mph
  C. 30 mph
  D. 45 mph

Answer A is incorrect. The programmable parameters page of the *Medium Heavy Composite Vehicle Type 3 Reference Booklet* says that the setting may be between 20 and 100 mph.

Answer B is correct. The programmable parameters page of the *Medium Heavy Composite Vehicle Type 3 Reference Booklet* says the setting is 25 mph.

Answer C is incorrect. A speed of 30 mph is an acceptable programmable parameter number; however, the current setting is 25 mph.

Answer D is incorrect. According to the booklet, 45 mph would be an acceptable setting; however, the booklet indicates the current setting is 25 mph.

24. Refer to the composite vehicle (diagram on page 39 of the booklet) to answer this question: The engine has "low injector current" active codes for injectors 1 and 2. Technician A says an open circuit at ECM connector 1 Terminal 6 could be the cause. Technician B says an open engine ground at G30 could be the cause. Who is correct?

TASK A.5

   A. A only

   B. B only

   C. Both A and B

   D. Neither A nor B

   Answer A is incorrect. An open circuit at ECM connector 1 terminal 6 would cause low current flow on injector 4 and prevent injector 4 from operating; it would not affect injectors 1 and 2.

   Answer B is incorrect. An open circuit at ECM ground at G30 could possibly prevent the engine from starting because the ECM would not have a complete path to ground; it would not affect only cylinders 1 and 2.

   Answer C is incorrect. Both Technicians are incorrect.

   Answer D is correct. Both Technicians are incorrect.

25. Refer to the composite vehicle (diagram on page 39 of the booklet) to answer the following question: While diagnosing an accelerator pedal position (APP) DTC using an oscilloscope, the technician finds a constant 5 volts at ECM connector 3, terminals 73 and 81, at all pedal positions. Which of the following could be the cause?

TASK B.5

   A. Open on ECM terminal 72

   B. Short to ground at ECM terminal 72

   C. Failed ECM

   D. Failed APP sensor

   Answer A is incorrect. An open at ECM terminal 72 would result in no voltage going to the APP; therefore, it would not be possible for the APP to send 5 volts out the signal wires.

   Answer B is incorrect. A short to ground at terminal 72 would prevent the APP from receiving 5 volts.

   Answer C is incorrect. There is no reason to suspect a failed ECM. The problem is that the APP is sending 5 volts to the ECM at all pedal positions.

   Answer D is correct. The APP has failed.

26. Refer to the composite vehicle to answer this question: The ECM has set an active DTC for "crankcase pressure sensor signal shorted low." When Connector BX is disconnected, the code becomes inactive. Technician A says the wiring harness is open. Technician B says the sensor is shorted. Who is correct?

TASK C.10

   A. A only

   B. B only

   C. Both A and B

   D. Neither A nor B

   Answer A is incorrect. If the DTC became inactive when the sensor was disconnected, then the sensor is the source of the DTC.

   Answer B is correct. Only Technician B is correct. The sensor is shorted. Disconnecting connector BX removed the short.

   Answer C is incorrect. Only Technician B is correct.

   Answer D is incorrect. Technician B is correct.

TASK D.7

27. Refer to the composite vehicle to answer the following question: All of the following are true concerning the fuel system on the composite vehicle EXCEPT:

A. The fuel transfer pump is mechanically driven.

B. The engine uses an EUI fuel system with six unit injectors.

C. All six injectors use a common power supply.

D. The fuel temperature sensor is a potentiometer.

Answer A is incorrect. The fuel transfer pump is a mechanically driven pump. Some other engines have an electrically driven pump.

Answer B is incorrect. The engine uses an EUI fuel system.

Answer C is correct. The engine uses three power supplies: one for injectors 1-2, one for injectors 3-4, and one for injectors 5-6.

Answer D is incorrect. The fuel temperature sensor is a thermister. A thermister changes resistance values based upon temperature.

TASK A.6

28. Refer to the composite vehicle to answer this question: While pulling a hill, the check engine light (CEL) and stop engine lamp (SEL) illuminate and the engine loses power. Technician A says engine coolant temperature above 230°F (110°C) could be the cause. Technician B says intake manifold temperature above 210°F (98.9°C) could be the cause. Who is correct?

A. A only

B. B only

C. Both A and B

D. Neither A nor B

Answer A is incorrect. Technician B is also correct.

Answer B is incorrect. Technician A is also correct.

Answer C is correct. Both Technicians are correct. The composite vehicle is programmed to derate and turn on the CEL and SEL for high engine, intake, and fuel temperatures.

Answer D is incorrect. Both Technicians are correct.

TASK B.7

29. Refer to the composite vehicle (diagram on page 40 of the booklet) to answer this question: The ambient air temperature sensor is multiplexed on which data bus?

A. J1587

B. J1708

C. J1922

D. J1939

Answer A is incorrect. The engine is equipped with a J1587 data bus, but it is not used for this signal.

Answer B is incorrect. Data bus J1708 is the protocol for J1587 and J1922.

Answer C is incorrect. The engine is not equipped with a J1922 data bus.

Answer D is correct. According to the reference booklet, the signal is sent out on the J1939 data bus.

30. Refer to the composite vehicle to answer this question: The composite vehicle has been in the shop multiple times for low power. All of the following could be the cause EXCEPT:

TASK C.11

   A. High EGR exhaust gas temperature.

   B. High coolant temperature.

   C. High coolant pressure.

   D. High DPF restriction.

   Answer A is incorrect. The ECM will derate for high EGR exhaust gas temperature; this can be caused by a restricted EGR cooler.

   Answer B is incorrect. The ECM will derate the engine for high coolant temperature. High coolant temperature can be caused by a restricted cooling system.

   Answer C is correct. The ECM is not programmed to reduce torque based upon coolant pressure; however, high exhaust backpressure can cause a derate condition.

   Answer D is incorrect. The ECM will derate the engine if high DPF restriction is detected. High exhaust restriction can be caused by a restricted DPF.

31. Refer to the composite vehicle to answer the following question: The truck will not accelerate above idle speed. Which of the following could be the cause?

TASK D.4

   A. One APP sensor signal has failed.

   B. A failed engine oil pressure sensor.

   C. Low coolant level.

   D. A failed cruise control switch.

   Answer A is correct. If one or both of the APP sensor signals fail, the engine will only idle.

   Answer B is incorrect. A failed engine oil pressure sensor could cause the engine to derate and shut down; however, it would not cause idle only.

   Answer C is incorrect. Low coolant level can cause the engine to derate and shut down; however, it will not cause an idle-only condition.

   Answer D is incorrect. A failed cruise control switch may cause the cruise control not to operate, but it will not cause an idle-only condition.

32. Refer to the composite vehicle to answer this question: A voltmeter is connected across terminals A and B of the clutch pedal switch. When the pedal is released, the voltmeter reads zero volts; when the pedal is depressed, the voltmeter reads 12 volts. Technician A says the switch is faulty. Technician B says the switch has an open ground. Who is correct?

TASK B.8

   A. A only

   B. B only

   C. Both A and B

   D. Neither A nor B

   Answer A is incorrect. The switch is showing correct voltage drop readings because the switch is normally a closed switch.

   Answer B is incorrect. If the ground was open, the voltmeter would indicate zero volts regardless of switch position.

   Answer C is incorrect. Neither Technician is correct.

   Answer D is correct. Neither Technician is correct.

TASK C.12

33. Refer to the composite vehicle for the following question: The composite diesel engine has air bubbles in the cooling system. The head gasket was replaced. After the test drive, the cooling system again has air in it. Which of the following would be the LEAST LIKELY cause?

A. Failure to fill the cooling system correctly

B. A failed EGR cooler

C. A cracked cylinder head

D. High exhaust backpressure

Answer A is incorrect. Care must be taken when filling the cooling system; filling too quickly can cause air pockets to be formed.

Answer B is incorrect. A failed EGR cooler can allow exhaust to enter the cooling system.

Answer C is incorrect. A cracked cylinder head can allow air in the cooling system, though this may not have been detected during the cylinder head gasket replacement.

Answer D is correct. High exhaust backpressure can cause low power, but would not introduce air into the cooling system.

TASK D.3

34. Refer to the composite vehicle to answer the following question: The composite diesel engine has low power. Which of the following is LEAST LIKELY to be the cause?

A. Cranking fuel pressure of 10 psi

B. 80 psi fuel pressure at rated RPM

C. Coolant temperature of 235°F (112.8°C)

D. Engine oil temperature of 245°F (118.3°C)

Answer A is incorrect. Cranking fuel pressure should be 20 psi. Low fuel pressure can cause low power.

Answer B is incorrect. Rated RPM fuel pressure should be between 90 and 100 psi. Low fuel pressure can cause low power.

Answer C is incorrect. The ECM will derate the engine when coolant temperature reaches 225°F. A coolant temperature of 235°F will result in a 60 percent derate.

Answer D is correct. The ECM will derate the engine when engine oil temperature reaches 260°F (126.7°C).

TASK B.6

35. Refer to the composite vehicle (diagram on page 39 of the booklet) to answer this question: Technician A says an open circuit 178 can cause a no-start condition. Technician B says an open circuit 171 can cause a no-start condition. Who is correct?

A. A only

B. B only

C. Both A and B

D. Neither A nor B

Answer A is correct. Only Technician A is correct. An open circuit 178 would prevent voltage from reaching the ECM; this could cause a no-start condition.

Answer B is incorrect. An open circuit 171 would prevent PTO operation from the dash-mounted switch. It would not, however, cause a no-start.

Answer C is incorrect. Technician B is incorrect.

Answer D is incorrect. Technician A is correct.

36. Refer to the composite vehicle to answer this question: Technician A says an engine cooling fan that runs all the time could be caused by a seized clutch. Technician B says an engine cooling fan that runs all the time could be caused by a faulty intake manifold pressure sensor. Who is correct?

TASK B.4

   A. A only
   B. B only
   C. Both A and B
   D. Neither A nor B

   Answer A is correct. Only Technician A is correct. A seized clutch will cause the fan to turn any time the pulley is turning.

   Answer B is incorrect. The composite diesel engine is not programmed to operate the fan in accordance with intake manifold pressure.

   Answer C is incorrect. Only Technician A is correct.

   Answer D is incorrect. Technician A is correct.

37. Refer to the composite vehicle (diagram on page 38 of the booklet) to answer this question: The engine has a DTC for EGR valve non-communication. Which of the following could be the cause?

TASK C.8

   A. A stuck open VGT
   B. A stuck closed EGR valve
   C. An open at connector JX pin 3
   D. A short at connector KX pin 1

   Answer A is incorrect. A stuck open VGT will not affect EGR communication, but may cause low power.

   Answer B is incorrect. A stuck closed EGR valve could set a code for incorrect EGR position, but it would not set an EGR communication fault.

   Answer C is correct. JX pin 3 is the connection for the EGR valve J1939 data bus. This could be the cause.

   Answer D is incorrect. KX pin 1 would affect the after treatment fuel shutoff valve, not the EGR valve.

38. Refer to the composite vehicle (diagram on page 37 of the booklet) to answer the following question: The composite vehicle has a misfire on cylinders 5 and 6. Any of the following could be the cause EXCEPT:

TASK D.2

   A. A leaking head gasket.
   B. An open on splice S11.
   C. A loose injector hold down on cylinder #4.
   D. An open on splice S13.

   Answer A is incorrect. A head gasket leaking between cylinders 5 and 6 can cause compression loss and a misfire on those two cylinders.

   Answer B is correct. An open on splice S11 would affect cylinders 1 and 2.

   Answer C is incorrect. A loose injector hold down on injector 4 can cause combustion chamber gases to enter the fuel stream. This will cause injectors 5 and 6 to misfire since they receive fuel after cylinder 4.

   Answer D is incorrect. An open on splice S13 would prevent voltage from the ECM to reach injectors 5 and 6; this could cause a misfire on cylinders 5 and 6.

TASK A.7

39. Refer to the composite vehicle to answer this question: Which of the following engine oil pressure conditions would cause the engine to derate, but not shut down?

    A.　7 psi @ 1000 rpm

    B.　18 psi @ 1000 rpm

    C.　20 psi @ 1400 rpm

    D.　10 psi @ 1400 rpm

A. Answer A is incorrect. The engine will shut down at 7 psi @ 1000 rpm. The engine will derate below 15 psi @ 1000 rpm, but will not shut down until the oil pressure drops to 8 psi @ 1000 rpm.

Answer B is incorrect. An oil pressure reading of 18 psi @ 1000 rpm will not cause a performance problem; derate starts at 15 psi when the engine is running at 1000 rpm.

Answer C is correct. An oil pressure reading of 20 psi @ 1400 rpm will cause the engine to derate, but not to shut down.

Answer D is incorrect. The engine is programmed to shut down when oil pressure is less than 12 psi @ 1400 rpm.

TASK B.6

40. Refer to the composite vehicle (diagram on page 39 of the booklet) to answer this question: Neither the SEL nor the CEL will illuminate in any ignition switch position. The engine operates normally otherwise. Which of the following could be the cause?

    A.　An open at ECM connector 3 pin 41

    B.　An open at ECM connector 3 pin 42

    C.　An open at splice S36

    D.　An open circuit 148

Answer A is incorrect. An open at Pin 41 would only affect the CEL.

Answer B is incorrect. An open at Pin 42 would only affect the SEL.

Answer C is correct. An open at splice S36 could prevent both bulbs from receiving the necessary 12 volts.

Answer D is incorrect. An open on circuit 148 would prevent the vehicle speed sensor (VSS) from reaching the ECM. It would not prevent the operation of the SEL or CEL.

TASK D.3

41. Refer to the composite vehicle to answer the following question: Which of the following is LEAST LIKELY to cause the CEL to be illuminated?

    A.　High coolant level

    B.　Low coolant level

    C.　Low oil level

    D.　High fuel temperature

Answer A is correct. High coolant level is not a parameter used to put the engine in protection mode, but low coolant level is.

Answer B is incorrect. Low coolant level can damage the engine and will put the ECM in protection mode.

Answer C is incorrect. Low oil level can damage an engine and will cause the ECM to enter into protection mode.

Answer D is incorrect. High fuel temperature can damage injector components, so the ECM will go into protection mode until the fuel temperature is reduced.

42. Refer to the composite vehicle (diagram on page 38 of the booklet) to answer this question: The EGR will not operate when commanded by the scan tool. The VGT works correctly. Which of the following could be the cause?

TASK B.7

   A. An open circuit 314

   B. An open circuit 313

   C. An open circuit 320

   D. An open circuit 315

Answer A is incorrect. Circuit 314 would affect operation of both the EGR and the VGT.

Answer B is incorrect. Circuit 313 would open the J1939 for the VGT and the EGR. This would affect the operation of both.

Answer C is incorrect. Circuit 320 could affect both, since it is the common ground for the VGT and the EGR.

Answer D is correct. Circuit 315 is the power supply to the EGR valve. This could cause the problem.

43. Refer to the composite vehicle to answer the following question: Technician A says the composite diesel engine is equipped with a returnless fuel system. Technician B says the fuel temperature sensor is located in the primary fuel filter. Who is correct?

TASK D.3

   A. A only

   B. B only

   C. Both A and B

   D. Neither A nor B

Answer A is incorrect. The fuel system has a return line; it is routed through the distribution block.

Answer B is incorrect. The fuel temperature sensor is located in the distribution block, not the primary filter.

Answer C is incorrect. Neither Technician is correct.

Answer D is correct. Neither Technician is correct.

44. Refer to the composite vehicle to answer this question: The engine cooling fan is running all the time. Technician A says a faulty intake manifold temperature sensor could be the cause. Technician B says a faulty coolant temperature sensor could be the cause. Who is correct?

TASK B.3

   A. A only

   B. B only

   C. Both A and B

   D. Neither A nor B

Answer A is incorrect. Technician B is also correct.

Answer B is incorrect. Technician A is also correct.

Answer C is correct. Both Technicians are correct. The engine cooling fan on the composite diesel engine is programmed to be engaged by the coolant temperature sensor or the intake manifold temperature sensor. So, if either sensor was sending a high temperature signal to the ECM, this could cause the concern.

Answer D is incorrect. Both Technicians are correct.

TASKS
D.1, D.3

45. Refer to the composite vehicle to answer the following question: The composite diesel engine will not start. There is no smoke from the exhaust while cranking. Technician A says this indicates no fuel is being delivered to the engine. Technician B says a faulty engine position sensor could be the cause of the no-start. Who is correct?

   A. A only
   B. B only
   C. Both A and B
   D. Neither A nor B

Answer A is incorrect. On older engines, the statement by Technician A would be true. However, the composite diesel engine has a DPF that catches the smoke.

Answer B is correct. Only Technician B is correct. If the engine position sensor failed, the ECM would not turn on the injectors. This would result in a no-start.

Answer C is incorrect. Only Technician B is correct.

Answer D is incorrect. Technician B is correct.

PREPARATION EXAM 3 – ANSWER KEY

1.	A	21.	B	41.	D
2.	A	22.	A	42.	B
3.	C	23.	D	43.	C
4.	A	24.	D	44.	A
5.	C	25.	A	45.	D
6.	D	26.	C		
7.	B	27.	A		
8.	D	28.	B		
9.	D	29.	C		
10.	C	30.	D		
11.	D	31.	C		
12.	D	32.	A		
13.	C	33.	D		
14.	A	34.	D		
15.	D	35.	C		
16.	B	36.	C		
17.	B	37.	A		
18.	B	38.	C		
19.	D	39.	D		
20.	C	40.	C		

PREPARATION EXAM 3 – EXPLANATIONS

1. An engine is diagnosed with low intake manifold pressure. The technician finds the turbocharger impeller severely worn. Which of the following could be the cause?

    A. A damaged air filter
    B. A stuck closed exhaust gas recirculation (EGR) valve
    C. A stuck open wastegate
    D. A failed after treatment fuel injector

TASKS
A.8, C.1

    Answer A is correct. A damaged air filter can allow dirt to enter the air stream and wear the impeller wheel, resulting in low intake manifold pressure.

    Answer B is incorrect. A stuck closed EGR valve would not cause low intake manifold pressure or a worn impeller wheel. It will cause higher than normal emissions.

    Answer C is incorrect. A stuck open wastegate could cause a turbocharger to be very slow building intake manifold pressure, but would not wear the impeller wheel.

    Answer D is incorrect. A failed after treatment injector could cause a damaged diesel oxidation catalyst (DOC) or diesel particulate filter (DPF), but would not cause a damaged impeller wheel.

TASK B.7

2. A diesel engine will not start. The technician finds an active diagnostic trouble code (DTC) for glow plugs 1, 3, 5, and 7. Which of the following is the LEAST LIKELY cause of the no-start condition?

A. A faulty ether injection solenoid valve

B. An open glow plug wiring harness

C. Low compression

D. A failed glow plug controller

Answer A is correct. Diesel engines that use glow plugs do not use ether injection; therefore, the failed ether injection solenoid would be the LEAST LIKELY.

Answer B is incorrect. An open glow plug harness on Bank 1 could cause an active DTC for glow plugs on cylinders 1, 3, 5, and 7.

Answer C is incorrect. Low compression can cause a no-start condition. An engine that has low compression will often have glow plugs that are burned open due to repeated start attempts.

Answer D is incorrect. A failed glow plug controller may prevent the operation of glow plugs 1, 3, 5, and 7, and cause a no-start condition.

TASK C.11

3. All of the following can cause a melted piston EXCEPT:

A. A broken exhaust rocker.

B. A stuck open injector.

C. A stuck closed injector.

D. A misdirected piston cooling nozzle.

Answer A is incorrect. A broken exhaust rocker will prevent the exhaust from escaping; this will prevent the heat from escaping the cylinder and will result in a melted piston.

Answer B is incorrect. A stuck open injector can dump excess fuel into the cylinder; this can result in a melted piston.

Answer C is correct. A stuck closed injector will not deliver fuel to the cylinder. This will cause the cylinder to be cold, but will not cause a melted piston.

Answer D is incorrect. A misdirected piston cooling nozzle can cause the piston to receive inadequate cooling oil spray, making it overheat and melt.

TASK B. 7

4. A truck equipped with an automated manual transmission will not go into gear. Technician A says a failed clutch switch could be the cause. Technician B says a failed cruise control switch could be the cause. Who is correct?

A. A only

B. B only

C. Both A and B

D. Neither A nor B

Answer A is correct. Only Technician A is correct. A failed clutch position switch would prevent the ECM from understanding true clutch position; therefore, it could not allow the transmission to shift into gear.

Answer B is incorrect. A failed cruise control switch would prevent cruise control operation, but it would not affect the transmission going into gear.

Answer C is incorrect. Only Technician A is correct.

Answer D is incorrect. Technician A is correct.

5. Technician A says a loose exhaust manifold can cause low intake manifold pressure. Technician B says a loose intake manifold can cause low intake manifold pressure. Who is correct?

**TASKS
A.9, C.1**

   A. A only

   B. B only

   C. Both A and B

   D. Neither A nor B

   Answer A is incorrect. Technician B is also correct.

   Answer B is incorrect. Technician A is also correct.

   Answer C is correct. Both Technicians are correct. A loose exhaust manifold will allow exhaust gas energy to escape prior to reaching the turbocharger. A loose intake manifold will allow the compressed air to escape. Both conditions will result in low intake manifold pressure.

   Answer D is incorrect. Both Technicians are correct.

6. Technician A says that a faulty APP can cause a no-start. Technician B says that an ambient air temperature sensor can cause a no-start. Who is correct?

TASK B.4

   A. A only

   B. B only

   C. Both A and B

   D. Neither A nor B

   Answer A is incorrect. A faulty APP will keep the engine from accelerating, but will not cause a no-start.

   Answer B is incorrect. A faulty ambient air temperature sensor may cause the engine to derate if it is reporting an overheat condition, but it will not cause a no-start condition.

   Answer C is incorrect. Neither Technician is correct.

   Answer D is correct. Neither Technician is correct.

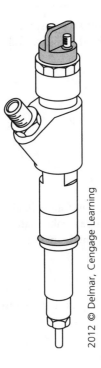

2012 © Delmar, Cengage Learning

**TASKS
D.5, D.9**

7. Referring to the figure above, the injector shown is being replaced. Which of the following is LEAST LIKELY to be performed as part of the repair?

A. The injector trim code may need to be installed in the ECM.

B. The injector rocker must be adjusted.

C. DTCs may need to be cleared from the ECM.

D. The engine should have a test run after injector replacement.

Answer A is incorrect. The new trim code usually needs to be installed during injector replacement. Failure to do so may result in a rough idle and/or poor cylinder performance.

Answer B is correct. This is a common rail injector. It does not have a rocker arm.

Answer C is incorrect. During any repair, the codes should be cleared to make sure they do not reappear during the test drive.

Answer D is incorrect. The engine must be test run after repairs to ensure the original concern was repaired and that no new concerns have arisen.

8. A diesel engine is making a knocking noise. The technician performs a cylinder cutout test using the scan tool. When cylinder #4 is deactivated, the knocking stops. Technician A says a failed vibration damper could be the cause. Technician B says a failed dual mass flywheel could be the cause. Who is correct?

TASK A.10

    A. A only
    B. B only
    C. Both A and B
    D. Neither A nor B

    Answer A is incorrect. A failed vibration damper can cause a knock. However, the knock will not be connected to a single cylinder.

    Answer B is incorrect. A failed dual mass flywheel will cause a knock; however, it will not be connected to a single cylinder. A knock that can be reduced by deactivating a cylinder is most likely caused by a worn rod bearing.

    Answer C is incorrect. Neither Technician is correct.

    Answer D is correct. Neither Technician is correct.

9. Technician A says the variable valve actuator system uses brake system air pressure to operate. Technician B says the variable valve actuator system only operates during engine brake operation. Who is correct?

TASK C.9

    A. A only
    B. B only
    C. Both A and B
    D. Neither A nor B

    Answer A is incorrect. The variable valve actuator system uses engine oil, not air pressure from the brake system, to operate.

    Answer B is incorrect. The variable valve actuator system opens the valves during engine operation to improve airflow and reduce emissions. It does not operate in conjunction with the engine brake.

    Answer C is incorrect. Neither Technician is correct.

    Answer D is correct. Neither Technician is correct.

10. The driver complains that the cruise control and speedometer do not operate on the truck. The antilock braking system (ABS) works correctly. Which of the following could be the cause?

TASK B.7

    A. An open J1939 data bus
    B. A shorted J1939 data bus
    C. An open vehicle speed sensor (VSS)
    D. A shorted cruise control switch

    Answer A is incorrect. If the J1939 data bus were open, the ABS would not work correctly.

    Answer B is incorrect. If the J1939 data bus were shorted, the ABS would not work correctly.

    Answer C is correct. An open VSS could cause the speedometer and cruise control to fail to function. The ABS would still work correctly because it uses the wheel speed sensors instead of the vehicle speed sensors.

    Answer D is incorrect. The speedometer would still work correctly even if the cruise control switch had a problem.

TASK C.12

11. An engine has had the turbocharger replaced due to a low boost pressure complaint. During the test drive after repairs, the technician finds the engine still has low boost. Which of the following is the LEAST LIKELY cause of the concern?

A. A restricted fuel filter

B. A restricted air filter

C. A restricted DPF

D. A restricted oil drain line

Answer A is incorrect. A restricted air filter will reduce fuel flow, decrease exhaust temperature, and reduce turbo boost.

Answer B is incorrect. A restricted air filter will reduce airflow and turbo boost.

Answer C is incorrect. A restricted DPF will reduce airflow and boost.

Answer D is correct. A restricted oil drain line will cause flooding of the turbo bearing housing and can cause the turbo to pass oil. It will not, however, cause the turbo to produce low boost pressure.

TASK B.6

12. The vehicle has the "OPT IDLE" lamp on the dash illuminated. Which of the following is indicated?

A. The vehicle will idle at a lower-than-normal RPM.

B. The remote throttle is engaged.

C. The APP has set a DTC and the vehicle will only idle.

D. The truck will shut off and restart automatically.

Answer A is incorrect. The truck may run at a higher-than-normal idle speed to help warm the engine.

Answer B is incorrect. The lamp is not connected to the remote throttle; it is the "Optimized Idle" lamp.

Answer C is incorrect. The CEL should be illuminated if there is an APP DTC.

Answer D is correct. The truck will shut off and restart automatically to maintain cab temperature and/or coolant temperature.

TASK C.8

13. The EGR system has an active DTC for the EGR airflow control (throttle) valve. Which of the following is the LEAST LIKELY cause of the code?

A. A binding throttle valve

B. A sticking throttle position sensor (TPS)

C. A sticking APP sensor

D. A shorted throttle valve wiring harness

Answer A is incorrect. A binding throttle valve can cause an increase in current flow and a nonresponsive valve. This will set a DTC.

Answer B is incorrect. A sticking TPS will result in a false TPS position being reported to the ECM, and a DTC.

Answer C is correct. A sticking APP would cause a false accelerator pedal position signal to be sent to the ECM. This would cause a DTC for the accelerator pedal, not the EGR airflow throttle.

Answer D is incorrect. A shorted wiring harness to the EGR throttle valve would set a DTC for the circuit.

14. Technician A says the cruise control switches can sometimes be used to change idle RPM. Technician B says the cruise control switches can be used to adjust maximum RPM. Who is correct?

TASK B.7

  A. A only

  B. B only

  C. Both A and B

  D. Neither A nor B

Answer A is correct. Only Technician A is correct. On some trucks, the cruise control switches can be used to increase or decrease idle RPM to help smooth engine operation.

Answer B is incorrect. Maximum RPM is a setting of the engine calibration and is not adjustable with the cruise control switches.

Answer C is incorrect. Only Technician A is correct.

Answer D is incorrect. Technician A is correct.

15. A vehicle equipped with a hydraulically actuated electronic unit injector (HEUI) fuel system idles poorly and has low power. Any of the following could be the cause EXCEPT:

TASK D.6

  A. The engine oil needs to be changed.

  B. The injectors are worn.

  C. The injector control pressure (ICP) sensor is faulty.

  D. The turbo charger wastegate is stuck open.

Answer A is incorrect. A HEUI fuel system uses engine oil under high pressure to operate the injectors. Engine oil that has had the additives depleted can cause the engine to idle roughly and have low power.

Answer B is incorrect. Worn injectors can cause smoke and poor performance.

Answer C is incorrect. A malfunctioning ICP can cause the ECM to control injection pressure erratically, resulting in rough idle and low power.

Answer D is correct. A stuck open wastegate will cause poor performance, but will not affect idle quality.

16. Technician A says the higher the resistance in an electrical circuit, the higher the amperage. Technician B says if voltage increases and resistance stays the same, that amperage will increase. Who is correct?

TASK B.8

  A. A only

  B. B only

  C. Both A and B

  D. Neither A nor B

Answer A is incorrect. As resistance increases, amperage decreases. The Ohm's Law formula is volts = (amps)(resistance).

Answer B is correct. Only Technician B is correct. If resistance is constant and voltage increases, then amperage will increase.

Answer C is incorrect. Only Technician B is correct.

Answer D is incorrect. Technician B is correct.

TASK C.10

17. Refer to the composite vehicle (diagram on page 38 of the booklet) to answer this question: All of the following can cause a crankcase ventilation system DTC EXCEPT:

    A.   An open circuit 302.

    B.   An open circuit 305.

    C.   A restricted crankcase filter.

    D.   A missing crankcase filter.

    Answer A is incorrect. Circuit 302 is the signal wire from the sensor. If it was open, it could cause a DTC.

    Answer B is correct. Circuit 305 is the signal wire from the EGR temperature sensor. It would cause a DTC for the EGR system, not the crankcase ventilation system.

    Answer C is incorrect. A restricted crankcase filter will cause abnormal crankcase pressure and set a DTC.

    Answer D is incorrect. A missing crankcase filter will cause abnormal crankcase pressure and will set a DTC.

TASK B.11

18. Refer to the composite vehicle engine to answer this question: The composite vehicle engine has set a DTC for the EGR pressure differential sensor. The technician replaces the sensor and clears the DTC. After the test drive, an active DTC for the EGR pressure differential sensor has reappeared. Which of the following is the most likely cause?

    A.   A faulty EGR pressure differential sensor

    B.   Restricted sensor hoses

    C.   A failed DPF

    D.   A failed DOC

    Answer A is incorrect. It is very unlikely that the new part is faulty. The most likely cause is that the original part was not faulty and the technician made a misdiagnosis.

    Answer B is correct. The hoses going to this sensor are prone to clogging. This is the most likely cause of the DTC.

    Answer C is incorrect. A failed DPF would set a DPF DTC, not an EGR DTC.

    Answer D is incorrect. A failed DOC would set a failed DOC DTC, not an EGR DTC.

TASK B.9

19. Refer to the composite vehicle (diagram on page 37 of the booklet) to answer this question: The engine brake is totally inoperative. Which of the following could be the cause?

    A.   An open circuit at ECM Connector 1 pin 9

    B.   An open circuit at ECM connecter 1 pin 10

    C.   An open circuit at ECM connector 1 pin 11

    D.   An open circuit at ECM connector 1 pin 12

    Answer A is incorrect. An open circuit at ECM connector 1 pin 9 would prevent injector 1 from having a ground and would cause a misfire on injector 1.

    Answer B is incorrect. An open circuit at ECM connector 1 pin 10 would not affect brake solenoid 3, only brake solenoids 1 and 2.

    Answer C is incorrect. An open circuit at ECM connector 1 pin 11 would affect only brake solenoid 3, not solenoids 1 and 2.

    Answer D is correct. An open circuit at ECM connector 1 pin 12 would open the ground to all three brake solenoids and prevent the engine brake from functioning.

20. Refer to the composite vehicle to answer this question: The check engine lamp (CEL) and stop engine lamp (SEL) illuminate and the engine shuts off 30 seconds later. The technician connects a scan tool and finds low oil pressure indicated in the data stream. A master gauge is connected and obtains an engine reading of 20 psi @ 600 rpm. The oil pressure rises as the engine RPM is increased and reaches 40 psi @ 2000 rpm. Which of the following could be the cause of the shutdown concern?

TASK A.7

   A. Worn main bearings
   B. A worn oil pump
   C. A faulty sending unit
   D. A stuck open pressure relief valve

Answer A is incorrect. Worn main bearings would cause low oil pressure. The master gauge indicates good oil pressure.

Answer B is incorrect. A worn oil pump would cause low oil pressure. The master gauge indicates good oil pressure.

Answer C is correct. If the sending unit was faulty, the electronic control module (ECM) could receive a false low oil pressure signal. That is what is occurring here. A faulty sending unit will send an inaccurate signal to the ECM. The master gauge indicates oil pressures are within specifications.

Answer D is incorrect. A stuck open pressure relief valve would cause low oil pressure. The master gauge indicates oil pressure well above specifications.

21. Refer to the composite vehicle to answer this question: Technician A says the CEL will start flashing 30 seconds prior to idle shutdown. Technician B says the operator can override the idle shutdown timer by depressing the service brake pedal. Who is correct?

TASK B.7

   A. A only
   B. B only
   C. Both A and B
   D. Neither A nor B

Answer A is incorrect. The SEL will start flashing 30 seconds prior to idle shutdown, not the CEL.

Answer B is correct. Only Technician B is correct. The driver can override the idle shutdown timer by depressing the clutch, service brake, or accelerator pedal.

Answer C is incorrect. Only Technician B is correct.

Answer D is incorrect. Technician B is correct.

TASK C.2

22. Refer to the composite vehicle to answer this question: Technician A says a faulty intake manifold temperature sensor can cause the engine cooling fan to stay on. Technician B says a faulty engine position sensor can cause the engine cooling fan to stay on. Who is correct?

    A.   A only

    B.   B only

    C.   Both A and B

    D.   Neither A nor B

Answer A is correct. Only Technician A is correct. The ECM is programmed to operate the cooling fan based upon the intake manifold temperature. A false high intake manifold temperature signal would cause the fan to operate constantly.

Answer B is incorrect. A faulty engine position sensor can set an active DTC and cause the SEL to illuminate, but would not cause the fan to operate constantly.

Answer C is incorrect. Only Technician A is correct.

Answer D is incorrect. Technician A is correct.

TASK B.9

23. Refer to the composite vehicle (diagram on page 40 of the booklet) to answer this question: There is an open circuit at connector SX pin 6. Which of the following would most likely be the result?

    A.   The ECM would start performing an active DPF regeneration.

    B.   The HETS lamp would start flashing.

    C.   The ECM would start performing a passive regeneration.

    D.   The SEL would illuminate.

Answer A is incorrect. The ECM cannot perform an active regeneration without a valid EGT3 signal.

Answer B is incorrect. The HETS lamp flashes for high EGT3 temperature.

Answer C is incorrect. The ECM will not perform a regeneration of the DPF if an accurate temperature at EGT3 cannot be determined.

Answer D is correct. An open circuit on the EGT sensor would set an active DTC and illuminate the SEL.

TASK D.4

24. Refer to the composite vehicle (diagram on page 39 of the booklet) to answer this question: Power take-off (PTO) mode works correctly from the remote PTO switch, but will not work from the cab-mounted PTO switch. Any of the following could be the cause EXCEPT:

    A.   An open circuit 170.

    B.   An EMC connector 3 pin 71 open.

    C.   A failed PTO switch.

    D.   An open circuit 171.

Answer A is incorrect. An open circuit 170 would prevent the ECM from receiving the PTO switch signal. This would cause the cab-mounted PTO to fail to operate.

Answer B is incorrect. ECM connector 3 pin 71 is the signal for the cab-mounted PTO. If it was open, the cab-mounted PTO would not work.

Answer C is incorrect. A cab-mounted PTO switch that was failed to open would prevent the cab-mounted PTO from working.

Answer D is correct. An open circuit 171 will prevent the remote PTO from operating correctly.

25. Refer to the composite vehicle to answer this question: The scan tool will communicate with the engine ECM, but will not communicate with the other modules on the truck. Which of the following could be the cause?

TASK B.7

A. An open circuit at connector XX

B. An open circuit 562

C. An open circuit 561

D. An open circuit at connector YX terminal B

Answer A is correct. An open circuit at connector XX would prevent the other modules on the J1939 data bus from communicating with the scan tool. However, the scan tool could still communicate on J1587.

Answer B is incorrect. An open circuit 562 would prevent J1587 communication.

Answer C is incorrect. An open circuit 561 would affect the after treatment DPF differential pressure sensor signal. It would not prevent data bus communications.

Answer D is incorrect. An open circuit at YX terminal B would prevent the scan tool from receiving Battery + from the data link connector and may prevent scan tool communication. It would not prevent the modules from communicating on the data bus.

26. Refer to the composite vehicle to answer this question: Injector #3 was replaced due to rough idle and poor performance during a cylinder power balance test. During the test drive, the technician finds the problem is still there. Which of the following is the LEAST LIKELY cause?

TASKS
D.7, D.8

A. Failure to reprogram the ECM with the new injector trim codes

B. Failure to set injector height correctly

C. A restricted primary fuel filter

D. A worn injector cam lobe

Answer A is incorrect. If the technician failed to enter the correct injector trim code during the replacement, the engine would still perform poorly.

Answer B is incorrect. The electronic unit injector (EUI) needs to be set at the correct height during installation. Failure to make this adjustment correctly can result in poor performance.

Answer C is correct. A restricted primary fuel filter would not affect only one cylinder during a cylinder power balance test.

Answer D is incorrect. A worn injector cam lobe can cause poor cylinder performance. This could be the cause of the concern.

2012 © Delmar, Cengage Learning

TASK B.2

27. Refer to the composite vehicle engine to answer this question: The lamp shown above is flashing. What does this indicate?

A. The exhaust gas temperature (EGT) 3 has indicated a temperature above 850°F (454.4°C).

B. The DPF delta pressure is higher than normal.

C. The coolant temperature is above 245°F (118.3°C).

D. The ECM has stored inactive codes.

Answer A is correct. This is the HEST lamp; it is illuminated when EGT3 is above 850° F.

Answer B is incorrect. When the DPF is restricted, the after treatment regeneration status lamp will be illuminated.

Answer C is incorrect. The ECM illuminates the CEL and SEL lamps when coolant temperature is high.

Answer D is incorrect. The CEL is illuminated when the ECM has stored inactive trouble codes.

TASK B.8

28. Refer to the composite vehicle to answer this question: Which of the following is the correct fuel injector resistance specification?

A. 0.05–0.5 ohms resistance

B. 0.5–5.0 ohms resistance

C. 5.0–50.0 ohms resistance

D. 50.0–500.0 ohms resistance

Answer A is incorrect. This is less than the correct specification. Resistance between 0.05 and 0.5 ohms indicates a shorted injector.

Answer B is correct. The correct specification is between 0.5 and 5.0 ohms resistance.

Answer C is incorrect. This is greater than the specification. Resistance between 5.0 and 50.0 ohms would indicate a failed injector.

Answer D is incorrect. This is excessive resistance; resistance between 50.0 and 500.0 ohms indicates a failed injector.

29. Refer to the composite vehicle to answer this question: A driver is concerned about low power and poor fuel economy. Technician A says a shorted engine cooling fan control switch will cause the engine cooling fan to operate continuously. Technician B says an engine cooling fan solenoid with excessive resistance can cause the engine cooling fan to operate continuously. Who is correct?

TASK D.8

    A. A only
    B. B only
    C. Both A and B
    D. Neither A nor B

Answer A is incorrect. Technician B is also correct. A shorted engine cooling fan switch would be the same as one that is closed. This tells the ECM to de-energize the engine cooling fan solenoid. This will cause the engine cooling fan to operate.

Answer B is incorrect. Technician A is also correct. A cooling fan solenoid with excessive resistance will not energize. If the solenoid does not energize, the fan will run continuously.

Answer C is correct. Both Technicians are correct.

Answer D is incorrect. Both Technicians are correct.

30. Refer to the composite vehicle to answer this question: Technician A says the vehicle speed sensor should have a resistance of between 100 and 1000 ohms. Technician B says the vehicle speed sensor is a three wire sensor. Who is correct?

TASK B.8

    A. A only
    B. B only
    C. Both A and B
    D. Neither A nor B

Answer A is incorrect. The correct specification is between 1100 and 1500 ohms resistance.

Answer B is incorrect. The vehicle speed sensor is a two wire sensor.

Answer C is incorrect. Neither Technician is correct.

Answer D is correct. Neither Technician is correct.

31. Refer to the composite vehicle to answer this question: Technician A says the coolant level sensor shares a ground with other switches. Technician B says the coolant level sensor connector is connector K. Who is correct?

TASK B.9

    A. A only
    B. B only
    C. Both A and B
    D. Neither A nor B

Answer A is incorrect. Technician B is also correct.

Answer B is incorrect. Technician A is also correct.

Answer C is correct. Both Technicians are correct. The coolant level sensor shares ECM connector 3 pin 77 as a common ground. K is indentified as the four wire connector on the coolant level sensor.

Answer D is incorrect. Both Technicians are correct.

**TASKS
B.9, B.8**

32. Refer to the composite vehicle to answer this question: The technician is measuring the J1939 backbone resistance with both resistors in place. Which of the following would indicate a proper resistance?

A.   60 ohms

B.   120 ohms

C.   240 ohms

D.   360 ohms

Answer A is correct. The J1939 data bus has two 120 ohm resistors in parallel; therefore, the correct measurement is 60 ohms.

Answer B is incorrect. Resistance of 120 ohms would indicate one resistor is missing.

Answer C is incorrect. Resistance of 240 ohms is 4 times the correct resistance and would indicate excess resistance in the circuit.

Answer D is incorrect. Resistance of 360 ohms is excessive and would likely cause no communication in the data bus.

TASK B.2

33. Refer to the composite vehicle (diagram on page 40 of the booklet) to answer the following question: The technician is troubleshooting a no communication on J1939 condition. Technician A says one of the J1939 backbone resistors is located in the variable geometry turbocharger (VGT) actuator. Technician B says the J1939 circuit is connected at the ECM at connector XX, and at Terminals A and B. Who is correct?

A.   A only

B.   B only

C.   Both A and B

D.   Neither A nor B

Answer A is incorrect. Although some manufacturers have installed one of the resistors in the VGT actuator, the composite engine is not wired this way. Both resistors are external.

Answer B is incorrect. The J1939 harness connects to the ECM at connector 4 pin 464 and 465.

Answer C is incorrect. Neither Technician is correct.

Answer D is correct. Neither Technician is correct.

TASK D.3

34. Refer to the composite vehicle to answer this question: Technician A says a restricted after treatment fuel injector can cause misfire on cylinder #5. Technician B says a restricted after treatment fuel injector can cause a low fuel system pressure.

A.   A only

B.   B only

C.   Both A and B

D.   Neither A nor B

Answer A is incorrect. A restricted after treatment fuel injector may prevent the DPF regeneration system from operating correctly; however, it will not cause a single cylinder misfire.

Answer B is incorrect. Low fuel system pressure may be caused by a restricted filter. It would not be caused by a restricted after treatment fuel injector.

Answer C is incorrect. Neither Technician is correct.

Answer D is correct. Neither Technician is correct.

35. Refer to the composite vehicle to answer this question: Technician A says active codes are indicated by the stop engine lamp (SEL) being illuminated. Technician B says inactive codes are indicated by the check engine light (CEL) being illuminated. Who is correct?

TASK B.6

   A. A only
   B. B only
   C. Both A and B
   D. Neither A nor B

Answer A is incorrect. Technician B is also correct.

Answer B is incorrect. Technician A is also correct.

Answer C is correct. Both Technicians are correct. The CEL indicates inactive codes and can be used in conjunction with the DPF lamp to indicate DPF status. The SEL indicates active codes and can be used in conjunction with the DPF lamp to indicate DPF status.

Answer D is incorrect. Both Technicians are correct.

36. Refer to the composite vehicle engine to answer this question: All of the following can cause the composite vehicle engine to shut off EXCEPT:

TASK B.2

   A. An idle shutdown timer.
   B. High coolant temperature.
   C. A restricted after treatment injector.
   D. Low oil pressure.

Answer A is incorrect. The idle shutdown timer can be set to shut down the engine after 1 to 100 minutes of idle time.

Answer B is incorrect. High coolant temperature above 240°F (115.6°C) can shut down the engine.

Answer C is correct. A restricted after treatment injector may reduce the spray from the injector, but would not cause the engine to shut down.

Answer D is incorrect. Oil pressure below the set pressure at a given RPM can cause shutdown.

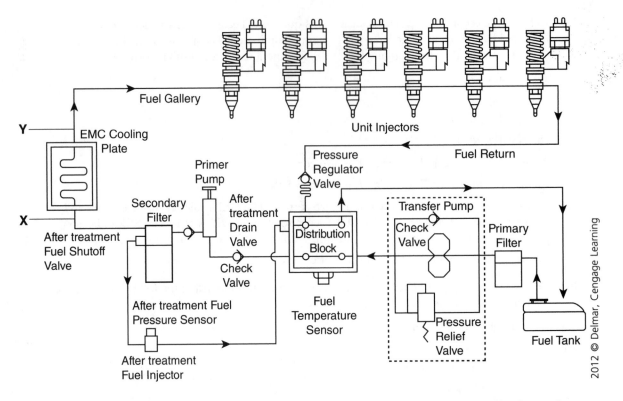

37. Refer to the composite vehicle engine to answer this question: Referring to the figure above, the composite vehicle engine has low power. With the engine at rated speed, Gauge X indicates 45 psi and Gauge Y indicates 44 psi. Which of the following is the LEAST LIKELY cause of low power?

TASKS B.10, D.3

A. A restricted ECM cooling plate

B. A restricted secondary filter

C. A worn transfer pump

D. A stuck open pressure regulator valve

Answer A is correct. A pressure drop of 1 psi across the ECM cooling plate is acceptable. The problem is that fuel pressure is much lower than specification and will result in low power.

Answer B is incorrect. A restricted secondary filter can cause low fuel pressure, which results in low power.

Answer C is incorrect. A worn transfer pump can cause low fuel pressure, and therefore low power.

Answer D is incorrect. A stuck open pressure regulating valve can cause low fuel pressure and low power.

38. Refer to the composite vehicle (diagram on page 39 of the booklet) to answer this question: The technician is diagnosing an APP1 DTC. While checking voltage at ECM Connector 3 Terminal 73, the technician finds 0.5 volts at 0 percent depressed throttle and 1.0 volts at 100 percent depressed throttle. Which of the following could be the cause of the DTC?

TASK B.2

    A. Circuit 172 is open.
    B. Circuit 173 is open.
    C. APP1 has failed.
    D. APP2 has failed.

    Answer A is incorrect. If circuit 172 were open, there would be no signal voltage at terminal 73.

    Answer B is incorrect. If circuit 173 were open, there would be no signal voltage at terminal 73.

    Answer C is correct. The signal voltage from APP1 is below specification. APP1 could have failed.

    Answer D is incorrect. The technician is diagnosing APP1, not APP2. The DTC is for APP1.

39. Refer to the composite vehicle to answer this question: The engine will not start. The technician finds 0.1 volts at ECM connector 3, terminal 80, key-on engine-off. While cranking, the voltage at ECM connector 3, terminal 80 rises to 5.5 volts. Which of the following is the LEAST LIKELY source of the no-start?

TASKS
B.6, B.8, B.9

    A. Circuit 180 has voltage drop.
    B. G30 is loose.
    C. S50 has corrosion.
    D. Circuit 179 has voltage drop.

    Answer A is incorrect. The test results indicate there could be voltage drop on circuit 180 between the ECM and S50, so this could be the cause.

    Answer B is incorrect. A loose connection at G30 could be the source of the voltage drop.

    Answer C is incorrect. Corrosion at S50 could be the source of the voltage drop; this would cause a no-start condition.

    Answer D is correct. Voltage drop is not being checked on circuit 179.

40. Refer to the composite vehicle to answer this question: Which of the following is an acceptable reading from the intake manifold pressure sensor?

TASK B.8

    A. 4.0 volts at 50 psig
    B. 3.3 volts at 37.5 psig
    C. 2.5 volts at 25 psig
    D. 1.5 volts at atmospheric pressure

    Answer A is incorrect. A reading of 5.0 volts is acceptable at 50 psig.

    Answer B is incorrect. A reading of 3.5 volts is acceptable at 37.5 psig.

    Answer C is correct. A reading of 2.5 volts is acceptable at 25 psig.

    Answer D is incorrect. A reading of 0.5 volts is acceptable at atmospheric pressure.

TASK B.10

41. Refer to the composite vehicle to answer this question: The composite diesel engine will not start. All of the following can cause the no-start EXCEPT:

A. ECM supply voltage of 8.0 volts.

B. ECM supply voltage of 17.0 volts.

C. Fuel injector resistance of 30.0 ohms each.

D. Fuel injector supply voltage of 120 volts.

Answer A is incorrect. The correct specification is a minimum of 9.0 volts.

Answer B is incorrect. The correct maximum specification is 16.0 volts.

Answer C is incorrect. Fuel injector resistance should be between 0.5 and 5.0 ohms each.

Answer D is correct. The fuel injector supply voltage should be between 100 and 120 volts.

TASK D.7

42. Refer to the composite vehicle to answer this question: Technician A says a fuel temperature sensor signal of 275°F (135°C) could indicate a sensor with high resistance. Technician B says that the fuel system is used to cool the engine ECM. Who is correct?

A. A only

B. B only

C. Both A and B

D. Neither A nor B

Answer A is incorrect. The fuel temperature sensor is a thermister. A high fuel temperature signal is caused by low resistance, not high resistance.

Answer B is correct. Only Technician B is correct. The fuel passes through an ECM cooling plate to cool the engine ECM.

Answer C is incorrect. Only Technician B is correct.

Answer D is incorrect. Technician B is correct.

TASK B.8

43. Refer to the composite vehicle to answer this question: Which of the following is an acceptable intake temperature sensor resistance?

A. 6–160 ohms

B. 600–16,000 ohms

C. 600–1600 ohms

D. 6000–16,000 ohms

Answer A is incorrect. A specification between 6 and 160 ohms is lower than specified in the composite vehicle booklet; resistance values in this range would set a DTC in the ECM.

Answer B is incorrect. A specification between 600 and 16,000 ohms is much wider than the specification in the book. Using this specification could result in a misdiagnosis.

Answer C is correct. The specification is between 600 and 1600 ohms.

Answer D is incorrect. A specification between 6000 and 16,000 ohms is too high. The correct specification is between 600 and 1600 ohms.

44. Refer to the composite vehicle to answer this question: Which of the following is LEAST LIKELY to cause a low power complaint?

    A. An open circuit 154

    B. An open circuit 16

    C. 45 psi fuel pressure at rated speed

    D. A faulty EGR Delta P sensor

TASK D.3

Answer A is correct. Circuit 154 would affect the cruise control; it would not affect power.

Answer B is incorrect. Circuit 16 being open would cause a misfire on cylinder #5 and low power.

Answer C is incorrect. Fuel pressure of 45 psi is less than the specification. This could cause low power.

Answer D is incorrect. The EGR Delta P sensor is used to determine EGR flow. If the sensor is faulty, the ECM can miscalculate EGR flow, causing low power.

45. Refer to the composite vehicle engine to answer this question: The driver complains that the idle shutdown timer will not function. Any of the following could be the problem EXCEPT:

    A. There is an active vehicle speed sensor fault.

    B. Ambient air temperature is 35°F (1.7°C).

    C. Engine coolant is 135°F (57.2°C).

    D. The service brake pedal switch is closed.

TASK B.10

Answer A is incorrect. If there is an active vehicle speed sensor fault, the idle shutdown timer is prevented from working for safety reasons.

Answer B is incorrect. Ambient air temperature must be between 40°F (4.4°C) and 80°F (26.7°C) for the idle shutdown timer to operate.

Answer C is incorrect. Engine coolant must be above 140°F (60.0°C) for the idle shutdown timer to operate.

Answer D is correct. The service brake pedal switch must be closed to allow the idle shutdown timer to operate.

PREPARATION EXAM 4 – ANSWER KEY

1.	C	21.	A	41.	D
2.	B	22.	A	42.	B
3.	C	23.	B	43.	B
4.	D	24.	D	44.	C
5.	A	25.	B	45.	B
6.	C	26.	D		
7.	D	27.	A		
8.	A	28.	A		
9.	D	29.	C		
10.	C	30.	D		
11.	D	31.	D		
12.	B	32.	C		
13.	C	33.	B		
14.	B	34.	B		
15.	B	35.	D		
16.	B	36.	D		
17.	D	37.	C		
18.	D	38.	B		
19.	C	39.	C		
20.	D	40.	D		

PREPARATION EXAM 4 – EXPLANATIONS

TASK B.6

1. A diesel engine cranks slowly and will not start. Technician A says low battery voltage may be the cause. Technician B says excessive voltage drop in the positive battery cables could be the cause. Who is correct?

    A. A only

    B. B only

    C. Both A and B

    D. Neither A nor B

    Answer A is incorrect. Technician B is also correct.

    Answer B is incorrect. Technician A is also correct.

    Answer C is correct. Both Technicians are correct. Low battery voltage, voltage drop in either the positive or negative battery cables, and/or a faulty starter motor can cause slow cranking speed and keep the engine from starting.

    Answer D is incorrect. Both Technicians are correct.

Cylinder 1	15 amps
Cylinder 2	15 amps
Cylinder 3	0 amp
Cylinder 4	0 amp
Cylinder 5	0 amp
Cylinder 6	0 amp
Cylinder 7	0 amp
Cylinder 8	0 amp

2. Referring to the table of diagnostic measurements above, a diesel engine equipped with glow plugs will not start in cold weather. A scan tool and amp clamp were used to measure individual glow plug amperage. Which of the following is the most likely cause of the hard to start condition?

TASK C.5

   A. An open glow plug timer

   B. Open glow plugs

   C. Low compression

   D. A faulty engine position sensor

Answer A is incorrect. None of the glow plugs will operate with an open glow plug timer.

Answer B is correct. Glow plugs for cylinders 3–8 have no current draw; they are most likely open and should be replaced.

Answer C is incorrect. Low compression can cause an engine to be hard to start in cold weather; the test results, however, indicate faulty glow plugs.

Answer D is incorrect. A faulty engine position sensor can cause a diesel engine to fail to start; the test results, however, indicate failed glow plugs.

3. The technician is uploading a new calibration file to the engine ECM while the ECM is still mounted on the engine. Which of the following would normally be part of this repair?

TASK B.3

   A. Remove the negative battery cable during the reflash.

   B. Remove the positive battery cable during the reflash.

   C. Connect a battery maintainer (low-rate charger) on the battery during the reflash.

   D. Connect a high-rate charger on the battery during the reflash.

Answer A is incorrect. When reflashing the ECM on the engine, the battery cables must be connected. This is how the ECM will receive battery voltage.

Answer B is incorrect. When reflashing the ECM on the truck both battery cables need to remain connected to the battery. If the battery cables were disconnected, the ECM would not receive voltage and the reflash could not occur.

Answer C is correct. A battery maintainer should be connected to the vehicle to ensure that battery voltage does not fall too low during the reflashing. If battery voltage falls too low, the ECM reflash may fail.

Answer D is incorrect. A high-rate charger should not be used; it may raise battery voltage too high during the reflash and cause the reflash to fail.

TASK B.6

4. Which of the following is LEAST LIKELY to cause a diesel engine to fail to start?

   A.   Low fuel pressure

   B.   Low fuel level

   C.   An open engine position sensor

   D.   An open vehicle speed sensor (VSS)

   Answer A is incorrect. Low fuel pressure can cause the engine to receive less fuel than is necessary to start.

   Answer B is incorrect. Low fuel level can starve the engine of fuel, resulting in a no-start.

   Answer C is incorrect. An open engine position sensor will not send the correct signal to the ECM; therefore, the ECM will not operate the injectors appropriately.

   Answer D is correct. An open VSS may prevent the cruise control from functioning properly, but will not prevent the engine from starting.

TASK C.3

5. The swinging vanes on a VGT are stuck. Which of the following will be the LEAST LIKELY result?

   A.   An active non-communication DTC for the VGT

   B.   An active position sensor DTC for the VGT

   C.   Reduced horsepower

   D.   Poor acceleration

   Answer A is correct. The VGT can still communicate even when the swinging vanes are stuck. The VGT will be unable to move the vanes, however.

   Answer B is incorrect. When the swinging vanes of the VGT are stuck, there normally is a DTC set for an error in the VGT position sensor. This is because the ECM has commanded the VGT to move the vanes and the position sensor did not reflect movement.

   Answer C is incorrect. Stuck swinging vanes can cause the boost pressure from the VGT to be lower than normal. This will reduce horsepower.

   Answer D is incorrect. Stuck swinging vanes can cause the boost pressure from the VGT to be lower than normal. This will result in poor acceleration.

TASK B.5

6. The technician is performing a stationary regeneration of the DPF. The following temperatures are recorded 15 minutes after the regeneration is started: EGT1 = 660°F (348.9°C), EGT2 = 900°F (482.2°C), EGT3 = 975°F (523.9°C). Which of the following is true?

   A.   EGT1 is higher than normal.

   B.   EGT2 is higher than normal.

   C.   All EGTs are normal.

   D.   EGT3 is higher than normal.

   Answer A is incorrect. EGT1 of 660°F is an approximately normal temperature. A temperature recording of 1000°F (537.8°C), would be higher than normal for EGT1.

   Answer B is incorrect. EGT2 is higher than EGT1; this indicates the DOC is functioning properly. An EGT2 of 1600°F (871.1°C) would be higher than normal.

   Answer C is correct. All EGTs are normal. EGT2 should be higher than EGT1 and EGT3 will be approximately the same as EGT2.

   Answer D is incorrect. EGT3 is about normal. An EGT3 of 1800°F (982.2°C) would be higher than normal.

7. Technician A says exhaust backpressure on trucks equipped with an exhaust after treatment device should be checked using a water manometer. Technician B says exhaust backpressure higher than normal can be caused by a missing DPF. Who is correct?

TASK C.4

   A. A only

   B. B only

   C. Both A and B

   D. Neither A nor B

Answer A is incorrect. Trucks with exhaust after treatment devices usually have higher backpressure; the pressure is measured using a mercury manometer or a psi gauge.

Answer B is incorrect. Higher-than-normal exhaust backpressure could be caused by a clogged DPF, not a missing DPF.

Answer C is incorrect. Neither Technician is correct.

Answer D is correct. Neither Technician is correct.

8. A diesel engine equipped with a hydraulically actuated electronically controlled unit injector (HEUI) fuel system will not start. Technician A says a faulty injection pressure regulator (IPR) can be the cause. Technician B says an open intake manifold pressure sensor can be the cause. Who is correct?

TASK D.6

   A. A only

   B. B only

   C. Both A and B

   D. Neither A nor B

Answer A is correct. Only Technician A is correct. A faulty IPR can result in very low injection actuation pressure, which can result in the injectors not delivering fuel.

Answer B is incorrect. An open intake manifold pressure sensor can cause low power or smoke; it will not cause a no-start condition.

Answer C is incorrect. Only Technician A is correct

Answer D is incorrect. Technician A is correct.

9. The driver complains that neither the cruise control nor the idle shutdown timer will work on a vehicle. Which of the following is the LEAST LIKELY cause?

TASK B.7

   A. A sticking treadle valve

   B. A faulty VSS

   C. A faulty accelerator pedal position (APP) sensor

   D. A stuck cruise control switch

Answer A is incorrect. A sticking treadle valve can cause the brakes to be applied. If the brake switch indicates the brakes are applied, the cruise control and the idle shutdown timer are both disabled.

Answer B is incorrect. Both the cruise control and the idle shutdown timer depend on a correct VSS signal. A faulty VSS signal could cause both to be inoperative.

Answer C is incorrect. The cruise control and the idle shutdown timer both rely on a correct signal from the APP sensor. A faulty sensor could cause both systems to fail to operate.

Answer D is correct. A stuck cruise control switch will affect cruise control, but not the idle shutdown timer.

TASK C.6

10. Technician A says a stationary regeneration of the DPF can be performed using a scan tool. Technician B says a stationary regeneration of the DPF can be performed using the dash-mounted switches. Who is correct?

    A. A only

    B. B only

    C. Both A and B

    D. Neither A nor B

    Answer A is incorrect. Technician B is also correct.

    Answer B is incorrect. Technician A is also correct.

    Answer C is correct. Both Technicians are correct. A stationary regeneration of the DPF can be performed with a scan tool or the dash-mounted switch, if the switch is provided. Some vehicles do not have a dash-mounted switch.

    Answer D is incorrect. Both Technicians are correct.

**TASKS B.6,
B.8**

11. The compression brake is weak on a diesel engine. The technician measures voltage at the engine brake solenoids and finds it to be within specification on all three. Which of the following is the LEAST LIKELY cause of the poor compression brake performance?

    A. Incorrect compression brake adjustment

    B. An open compression brake solenoid

    C. A leaking master piston seal

    D. An open brake selector switch

    Answer A is incorrect. Incorrect compression brake adjustment can cause a weak compression brake.

    Answer B is incorrect. An open solenoid can cause poor brake performance because only four of the six cylinders would be braking.

    Answer C is incorrect. A leaking master piston seal would cause the brake to fail to open the exhaust valve correctly, resulting in poor brake performance.

    Answer D is correct. If the brake selector switch failed open, there would be zero voltage at all three solenoids.

TASK C.7

12. The cooling system will not hold pressure during a cooling system pressure test. Which of the following could be the cause?

    A. A leaking ECM cooler

    B. A leaking EGR cooler

    C. A leaking charge air cooler

    D. A leaking power steering cooler

    Answer A is incorrect. The ECM cooler has fuel passages to cool the ECM. Engine coolant is not in this cooler.

    Answer B is correct. The EGR cooler is a coolant-based cooler; this could be the cause of the cooling system's failure to hold pressure.

    Answer C is incorrect. The charge air cooler is an air-to-air cooler. It would not cause a cooling system leak. It should be noted, however, that some trucks use a jacket water aftercooler that would allow a cooling system leak.

    Answer D is incorrect. The power steering cooler does not use coolant.

13. Technician A says a damaged DPF can cause low power. Technician B says a damaged DOC can cause low power. Who is correct?

TASK B.1

    A. A only

    B. B only

    C. Both A and B

    D. Neither A nor B

Answer A is incorrect. Technician B is also correct.

Answer B is incorrect. Technician A is also correct.

Answer C is correct. Both Technicians are correct. If the DOC or DPF is damaged, it can cause restricted exhaust. Restricted exhaust can cause low power.

Answer D is incorrect. Both Technicians are correct.

14. A diesel engine will not start. The technician disconnects the 5 volt reference wire for the pressure and temperature sensors from the ECM and the engine starts. Technician A says the ECM is faulty. Technician B says there is a short in the 5 volt reference circuit. Who is correct?

TASK D.8

    A. A only

    B. B only

    C. Both A and B

    D. Neither A nor B

Answer A is incorrect. If the engine will start after the wire is disconnected from the ECM, the ECM is not faulty. Instead, the wire is shorted, causing the ECM to enter protection mode.

Answer B is correct. Only Technician B is correct. A shorted circuit causes increased current draw. Some ECMs will totally shut down to protect the ECM from overload.

Answer C is incorrect. Only Technician B is correct.

Answer D is incorrect. Technician B is correct.

15. A diesel engine will not start. Which of the following could be the cause?

TASK C.6

    A. A faulty DPF differential pressure sensor

    B. A failed engine position sensor

    C. A stuck closed after treatment injector

    D. A stuck closed after treatment fuel shutoff valve

Answer A is incorrect. A failed DPF differential pressure sensor will prohibit DPF regeneration, but will not prohibit engine starting.

Answer B is correct. A failed engine position sensor cannot send a signal to the ECM, so the ECM will not energize the fuel injectors.

Answer C is incorrect. A stuck closed after treatment injector will prevent active regeneration of the DPF, but will not prevent engine starting.

Answer D is incorrect. A stuck closed after treatment fuel shutoff valve will prevent active regeneration of the DPF, but will not prevent engine starting.

TASK B.2

16. The ECM has active DTCs for the engine oil temperature sensor, the ambient air temperature sensor, and the intake manifold temperature sensor. The freeze frame data indicates that the DTCs were all set at the same time. Which of the following could be the cause?

    A.   An open engine oil temperature sensor
    B.   An open common 5 volt reference wire
    C.   An open signal wire from the intake manifold temperature sensor
    D.   An overheated vehicle

Answer A is incorrect. An open oil temperature sensor would not set a DTC for the other sensors.

Answer B is correct. Often these sensors will share a common supply wire or a common ground wire. An open in the common wire would set a DTC for all the sensors sharing that wire.

Answer C is incorrect. An open signal wire from one sensor would not cause the other sensors to fail.

Answer D is incorrect. If the vehicle was overheated, DTCs would likely be set; however, the ambient air temperature would not be affected.

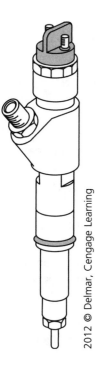

2012 © Delmar, Cengage Learning

17. Referring to the figure above, Technician A says the injector shown uses engine oil to create high-pressure fuel injection. Technician B says the injector shown above uses a camshaft lobe and rocker arm to create high pressure. Who is correct?

TASK D.5

A. A only

B. B only

C. Both A and B

D. Neither A nor B

Answer A is incorrect. The injector shown above is a high-pressure common rail (HPCR) injector. It receives fuel under high pressure from a high-pressure fuel pump. It does not use engine oil to create injection pressure. The HEUI injector uses high pressure oil to create the pressure needed to inject the diesel fuel.

Answer B is incorrect. The injector shown above is a high-pressure common rail (HPCR) injector. It receives fuel under high pressure from a high-pressure pump. An EUI injector uses a camshaft lobe and rocker arm to create the high pressure for fuel injection.

Answer C is incorrect. Neither Technician is correct.

Answer D is correct. Neither Technician is correct.

TASK A.11

18. Refer to the composite vehicle to answer this question: The customer complains that the speedometer reads incorrectly. Any of the following could be the cause EXCEPT:

    A. Tire revolutions per mile set at 600.

    B. Tail shaft teeth set at 18.

    C. Rear axle ratio set at 5:1.

    D. Maximum engine speed without vehicle speed sensor (VSS) set at 2000 rpm.

    Answer A is incorrect. The tire revolutions per mile should be set at 501. A setting of 600 could result in an incorrect speedometer reading.

    Answer B is incorrect. The tail shaft teeth should be set at 16. A setting of 18 could result in an incorrect speedometer reading.

    Answer C is incorrect. The rear axle ratio should be set at 4:1. A setting of 5:1 could result in an incorrect speedometer reading.

    Answer D is correct. Setting maximum engine speed at 2000 rpm would allow higher than specified engine RPM. If the VSS failed, it would not result in an incorrect speedometer reading.

**TASKS
B.1, B.6**

19. Refer to the composite engine to answer this question: The ECM has set a diagnostic trouble code (DTC) for a nonresponsive VGT actuator motor. Which of the following is the LEAST LIKELY cause?

    A. A binding VGT actuator arm

    B. A sticking VGT nozzle ring

    C. Sticking VGT swinging vanes

    D. A stuck VGT motor

    Answer A is incorrect. A binding actuator arm can prevent the sliding nozzle from operating and will set a DTC for the motor.

    Answer B is incorrect. There are two main VGT styles currently being used, sliding nozzle and swinging vanes. The composite engine uses the sliding nozzle design.

    Answer C is correct. The VGT on the composite engine uses a sliding nozzle ring, not swinging vanes.

    Answer D is incorrect. A stuck VGT motor will prevent movement and set this DTC.

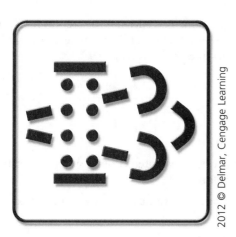

20. Refer to the composite engine to answer this question: The SEL and the light shown above are illuminated on the dash. What is indicated?

TASK B.2

A.   Level 1 soot load

B.   Level 2 soot load

C.   Level 3 soot load

D.   Level 4 soot load

Answer A is incorrect. At level 1 soot load the DPF status lamp will be illuminated.

Answer B is incorrect. At level 2 soot load the DPF status lamp will be flashing.

Answer C is incorrect. At level 3 soot load the DPF status lamp is flashing and the CEL is illuminated.

Answer D is correct. At level 4 soot load the DPF status lamp is flashing and the SEL is illuminated.

21. Refer to the composite engine to answer this question: All of the following could cause a misfire on cylinder #3 of the composite diesel engine EXCEPT:

TASK D.8

A.   A restricted ECM cooling plate.

B.   A broken injector spring.

C.   A worn intake cam lobe.

D.   A worn exhaust cam lobe.

Answer A is correct. A restricted ECM cooling plate will cause low power, but will not cause a single cylinder misfire.

Answer B is incorrect. A broken injector spring on cylinder #3 would cause the injector to short stroke and deliver less than normal fuel.

Answer C is incorrect. A worn intake cam lobe will reduce the amount of fresh air entering the combustion chamber. This will reduce power and can cause a single cylinder misfire.

Answer D is incorrect. A worn exhaust cam lobe will prevent the exhaust gases from escaping and prevent the cylinder from filling with fresh air on the intake stroke. This can cause a single cylinder misfire.

TASK B.10

22. Refer to the composite vehicle to answer this question: Technician A says repeat head gasket failures may be caused by a restricted cooling system. Technician B says repeat head gasket failures can be caused by a restricted primary fuel filter. Who is correct?

   A. A only

   B. B only

   C. Both A and B

   D. Neither A nor B

   Answer A is correct. Only Technician A is correct. A restricted cooling system can cause high operating temperatures and head gasket failure.

   Answer B is incorrect. A restricted primary fuel filter can cause low power, but will not cause a head gasket failure.

   Answer C is incorrect. Only Technician A is correct.

   Answer D is incorrect. Technician A is correct.

TASK B.5

23. Refer to the composite vehicle to answer this question: The diesel engine will not perform a stationary regeneration of the DPF. Technician A says a restricted EGR passage could be the cause. Technician B says a faulty after treatment injector could be the cause. Who is correct?

   A. A only

   B. B only

   C. Both A and B

   D. Neither A nor B

   Answer A is incorrect. A restricted EGR passage may prevent sufficient EGR flow; however, the EGR is not necessary during DPF regeneration.

   Answer B is correct. Only Technician B is correct. The after treatment fuel injector sprays fuel into the exhaust stream to heat the exhaust to initiate regeneration of the DPF. If the injector is faulty, the fuel may not be sprayed into the exhaust stream.

   Answer C is incorrect. Only Technician B is correct.

   Answer D is incorrect. Technician B is correct.

TASK B.5

24. Refer to the composite vehicle to answer this question: The vehicle has set a DTC for EGR valve position. During diagnosis, the technician finds that the EGR valve position sensor voltage is 2.5 volts, regardless of engine RPM or engine load. Technician A says the EGR valve is stuck fully shut. Technician B says the EGR valve J1939 data bus circuit is open. Who is correct?

   A. A only

   B. B only

   C. Both A and B

   D. Neither A nor B

   Answer A is incorrect. If the EGR valve were stuck fully shut, the sensor voltage would be 0.5 volts.

   Answer B is incorrect. If the EGR valve J1939 data bus circuit were open, the EGR valve position sensor voltage would not be reported to the ECM.

   Answer C is incorrect. Neither Technician is correct. The EGR valve should move in relation to engine speed and load. It did not, so it must be stuck, and 2.5 volts is the 50 percent (one-half) position.

   Answer D is correct. Neither Technician is correct. A constant voltage reading of 2.5 volts indicates the valve is stuck at 50 percent.

25. Refer to the composite vehicle to answer this question: The composite vehicle is being used in a refuse operation. The technician finds the diesel particulate filter (DPF) must have a stationary regeneration performed using a scan tool every two days. Any of following could be the cause EXCEPT:

TASK A.12

 A. Mobile/active regeneration has been disabled.

 B. The high exhaust temperature lamp has been disabled.

 C. The DPF regeneration permit switch has been disabled.

 D. Stationary regeneration in power take-off (PTO) mode has been disabled.

Answer A is incorrect. If the mobile/active regeneration were disabled, it would prevent the electronic control module (ECM) from performing a regeneration when the vehicle was moving. This could cause the need to perform a stationary regeneration every two days.

Answer B is correct. If the lamp were disabled, it would prevent the ECM from illuminating the lamp when exhaust temperature is high. It would not prevent regeneration.

Answer C is incorrect. If the permit switch were disabled, it would prevent the ECM from performing the regeneration. This could cause the need for stationary regeneration.

Answer D is incorrect. If the ECM cannot perform a regeneration in PTO mode, this could cause the need for stationary regeneration every two days.

26. Refer to the composite vehicle to answer this question: The engine cooling fan will not engage using the dash-mounted switch. It works correctly when commanded by the scan tool. Technician A says the switch may be stuck closed. Technician B says a failed engine cooling fan solenoid could be the cause. Who is correct?

TASK B.4

 A. A only

 B. B only

 C. Both A and B

 D. Neither A nor B

Answer A is incorrect. A stuck closed switch will cause the fan to run continuously.

Answer B is incorrect. If the solenoid had failed, the clutch would not work correctly when commanded by the scan tool.

Answer C is incorrect. Neither Technician is correct.

Answer D is correct. Neither Technician is correct.

27. Refer to the composite vehicle (diagram on page 37 of the booklet) to answer this question: Any of the following could cause the engine brake to function poorly EXCEPT:

TASK B.9

 A. An open circuit at ECM connector 1 pin 9.

 B. An open circuit at ECM connecter 1 pin 10.

 C. An open circuit at ECM connector 1 pin 11.

 D. An open circuit at ECM connector 1 pin 12.

Answer A is correct. An open circuit at ECM connector 1 pin 9 would prevent injector 1 from having a ground and would cause a misfire on injector 1, but would not affect engine brake functionality.

Answer B is incorrect. An open circuit at ECM connector 1 pin 10 would affect the engine brake solenoid for cylinders 1 and 2 and the solenoid for cylinders 5 and 6.

Answer C is incorrect. An open circuit at ECM connector 1 pin 11 would affect the engine brake solenoid for cylinders 3 and 4.

Answer D is incorrect. An open circuit at ECM connector 1 pin 12 would open the ground to all three solenoids and prevent the engine brake from functioning.

Cylinder 1	100 rpm
Cylinder 2	100 rpm
Cylinder 3	10 rpm
Cylinder 4	10 rpm
Cylinder 5	100 rpm
Cylinder 6	100 rpm

TASK D.1

28. Refer to the composite vehicle to answer this question: The test results in the table above are from a power balance test on the composite engine. Which of the following is the most likely cause?

    A.   A leaking cylinder head gasket

    B.   A leaking EGR cooler gasket

    C.   An open at splice S11

    D.   An open at splice S13

Answer A is correct. When two adjacent cylinders are low on power, a leaking head gasket is suspect. A compression test will help isolate this condition.

Answer B is incorrect. A leaking EGR cooler will cause a coolant loss; it will not, however, normally cause two adjacent cylinders to have low power.

Answer C is incorrect. An open at splice S11 would cause low test results on cylinders 1 and 2.

Answer D is incorrect. An open at splice S13 would cause low test results on cylinders 5 and 6.

TASK B.2

29. Refer to the composite vehicle to answer this question: The diagnostic switch has been pressed. The SEL and CEL both momentarily turn on and then off. Which of the following could be the cause?

    A.   There are multiple active fault codes.

    B.   There are multiple inactive fault codes.

    C.   There are no fault codes.

    D.   The ECM is malfunctioning.

Answer A is incorrect. Active fault codes are displayed on the SEL.

Answer B is incorrect. Inactive fault codes are displayed on the CEL.

Answer C is correct. This is how the system should respond if there are no active or inactive fault codes.

Answer D is incorrect. This is a normal condition and does not indicate a failure.

30. Refer to the composite vehicle to answer this question: All of the following are programmable parameters on the composite vehicle EXCEPT:

TASK B.3

A. Idle shutdown manual override.

B. Minimum fan on-time.

C. Tire size.

D. Steering wheel diameter.

Answer A is incorrect. The idle shutdown manual override is programmable to allow the owner to disable the idle shutdown timer.

Answer B is incorrect. Minimum fan on-time is programmed to help eliminate excessive cycling of the fan clutch.

Answer C is incorrect. Tire size is programmed for use by the ECM for speedometer and odometer information.

Answer D is correct. Steering wheel diameter will vary from truck to truck; the normal size is 22 inches. This information is not stored in the ECM.

31. Refer to the composite vehicle to answer this question: Technician A says the ambient air temperature sensor is directly wired to the engine ECM. Technician B says the ambient air temperature sensor voltage should be 3.0 volts at 90°F (32.2°C). Who is correct?

TASK B.8

A. A only

B. B only

C. Both A and B

D. Neither A nor B

Answer A is incorrect. The ambient air temperature sensor is indirectly wired to the engine ECM through the BCM, and the information is transmitted through the data bus.

Answer B is incorrect. The sensor signal voltage should be 2.5 volts at 90°F.

Answer C is incorrect. Neither Technician is correct.

Answer D is correct. Neither Technician is correct.

32. Refer to the composite vehicle to answer this question: Technician A says a failed coolant temperature sensor can prevent the regeneration of the DPF. Technician B says a failed coolant temperature sensor can cause the instrument panel temperature gauge to read incorrectly. Who is correct?

TASK A.6

A. A only

B. B only

C. Both A and B

D. Neither A nor B

Answer A is incorrect. Technician B is also correct.

Answer B is incorrect. Technician A is also correct.

Answer C is correct. Both Technicians are correct. The coolant temperature sensor signal is used for the instrument panel gauge, DPF regeneration, and cooling fan operation.

Answer D is incorrect. Both Technicians are correct.

TASK B.3

33. Refer to the composite vehicle to answer this question: The fan control A/C pressure switch programmable parameter has been set to disabled. Which of the following is LEAST LIKELY to occur?

A. The A/C will operate poorly at low vehicle speeds.

B. The A/C will operate poorly at highway speeds.

C. The A/C high-side pressure will be higher than normal.

D. The A/C pressure switch will open.

Answer A is incorrect. The fan is needed to help keep the condenser cool during low-speed operation.

Answer B is correct. The fan is not needed at highway speeds to keep the condenser cool.

Answer C is incorrect. If the fan does not operate when needed to keep the condenser cool, the A/C high-side pressure will be higher than normal.

Answer D is incorrect. The A/C pressure switch will open if the high-side pressure climbs too high. This could happen at low vehicle speeds if the parameter is disabled.

TASK B.2

34. Refer to the composite vehicle (diagram on page 40 of the booklet) to answer this question: The ECM has set an active DTC for EGT1 temperature below normal. Which of the following could be the cause?

A. An open connector SX

B. An open circuit 555

C. An open circuit at ECM pin 456

D. A short-to-ground on circuit 556

Answer A is incorrect. An open SX connector would set a code for all three EGT sensors.

Answer B is correct. Circuit 555 is the signal wire for EGT1. This could cause the active DTC.

Answer C is incorrect. An open at ECM pin 456 would set a code for all three EGT sensors.

Answer D is incorrect. Circuit 556 is a ground circuit; it should be grounded.

Fuel temperature sensor connected	0.0 volts
Fuel temperature sensor disconnected and shorted	5.0 volts

35. Refer to the composite vehicle to answer this question: There is an active DTC for the fuel temperature sensor. The technician connects the scan tool and monitors the fuel temperature sensor voltage while connecting and disconnecting the fuel temperature sensor. The test results are listed in the table above. Which of the following is the cause of the active DTC?

    TASK D.1

    A. The fuel temperature is higher than normal.

    B. The fuel temperature is lower than normal.

    C. The fuel temperature sensor is open.

    D. The fuel temperature sensor is shorted.

Answer A is incorrect. These test results indicate a problem with the temperature sensor circuit, not the fuel temperature itself.

Answer B is incorrect. If the fuel temperature were lower than normal, the voltage would not be 0.0 volts and 5.0 volts.

Answer C is incorrect. If the fuel temperature sensor were open, the voltage would not change when the sensor was disconnected.

Answer D is correct. These test results indicate a shorted sensor because both the highest and lowest voltages possible are displayed when the technician performed the test.

36. Refer to the composite vehicle to answer this question: All of the following are true concerning the exhaust gas temperature (EGT) sensors EXCEPT:

    A. EGT 1 is also referred to as the after treatment diesel oxidation catalyst (DOC) inlet temperature sensor.

    TASK B.6

    B. EGT 2 is also referred to as the after treatment DPF inlet temperature sensor.

    C. EGT 3 is also referred to as the after treatment DPF outlet temperature sensor.

    D. An overheated EGT 1 will illuminate the high exhaust system temperature (HEST) lamp.

Answer A is incorrect. EGT 1 is located in front of the after treatment device and is also referred to as the after treatment DOC inlet temperature sensor.

Answer B is incorrect. EGT 2 is located in the middle of the after treatment device and is also referred to as the after treatment DPF inlet temperature sensor.

Answer C is incorrect. EGT 3 is located at the outlet device and is also referred to as the after treatment DPF outlet temperature sensor.

Answer D is correct. EGT3 is the signal used to illuminate the HEST lamp, not EGT1.

37. Refer to the composite vehicle to answer this question: The idle shutdown timer does not shut the vehicle off after the programmed time. Technician A says cold ambient temperatures could be the cause. Technician B says high ambient temperatures could be the cause. Who is correct?

    A. A only

    TASK B.7

    B. B only

    C. Both A and B

    D. Neither A nor B

Answer A is incorrect. Technician B is also correct.

Answer B is incorrect. Technician A is also correct.

Answer C is correct. Both Technicians are correct. The idle shutdown timer is programmed to operate only between the ambient temperatures of 40°F (4.4°C) and 80°F (26.7°C).

Answer D is incorrect. Both Technicians are correct.

Accelerator pedal position (APP)	0%
Engine speed	200 rpm
Camshaft position sensor/Engine position signal 1 (CMP/EPS1)	No
Crankshaft position sensor/Engine position signal 2 (CKP/EPS2)	Yes
Engine coolant level	Normal

TASK B.5

38. Refer to the composite vehicle to answer this question: The composite diesel engine will not start. The scan tool data shown in the table above is retrieved while cranking the engine. Which of the following could be the cause?

   A. A faulty APP sensor

   B. A faulty EPS1

   C. A faulty EPS2

   D. Low coolant level

Answer A is incorrect. An APP reading of 0 percent would be a normal signal and would not prevent the engine from starting.

Answer B is correct. A faulty CMP/EPS1 can be the cause. The scan tool data shows "No." The ECM must see both EPS1 and EPS2 to start the engine.

Answer C is incorrect. A faulty CKP/EPS2 is not indicated in the scan tool data. The signal shows "Yes" and engine RPM is displayed.

Answer D is incorrect. The engine coolant level shows "Normal." This would not prevent the engine from starting.

TASK A.14

39. Refer to the composite engine (diagram on page 38 of the booklet) to answer this question: Technician A says wiring harness connector IX will be located on the exhaust gas recirculation (EGR) valve motor. Technician B says wiring harness connector JX will be a pigtail connector on the variable geometry turbocharge (VGT) actuator. Who is correct?

   A. A only

   B. B only

   C. Both A and B

   D. Neither A nor B

Answer A is incorrect. Technician B is also correct.

Answer B is incorrect. Technician A is also correct.

Answer C is correct. Both Technicians are correct. Referring to the wiring diagram for the composite engine, connector IX is shown as connecting directly to the EGR valve motor and connector JX is shown as being a pigtail connector on the VGT actuator.

Answer D is incorrect. Both Technicians are correct.

40. Refer to the composite vehicle to answer this question: Which of the following signals would be considered out of value range?

    A. After treatment DOC inlet temperature of 1000°F (537.8°C)

    B. After treatment DPF differential pressure of 28 in. Hg

    C. After treatment DPF outlet temperature of 1400°F (760°C)

    D. After treatment fuel pressure of 300 psi

TASK B.5

    Answer A is incorrect. The normal value range for the after treatment DOC inlet temperature is between 0°F and 2000°F (−17.8°C and 1093.3°C).

    Answer B is incorrect. The normal value range for the after treatment DPF differential pressure is between 0 and 30 in. Hg.

    Answer C is incorrect. The normal value range for after treatment DPF outlet temperature is between 0°F and 2000°F.

    Answer D is correct. The normal value range for the after treatment fuel pressure is between 0 psi and 200 psi.

41. Refer to the composite vehicle to answer this question: The engine oil pressure instrument panel gauge indicates low oil pressure. The scan tool indicates normal oil pressure. Technician A says the oil pressure sensor may be faulty. Technician B says there may be high resistance on ECM circuit 118. Who is correct?

    A. A only

    B. B only

    C. Both A and B

    D. Neither A nor B

TASKS
B.5, B.7

    Answer A is incorrect. If the oil pressure sensor were faulty, the gauge and the scan tool would still indicate the same pressure. Both devices use information from the engine oil pressure sensor.

    Answer B is incorrect. If there were high resistance on circuit 188, both the gauge and the scan tool would still indicate the same pressure. The most likely cause of this problem is a faulty gauge.

    Answer C is incorrect. Neither Technician is correct.

    Answer D is correct. Neither Technician is correct.

42. Refer to the composite vehicle to answer this question: All of the following are true concerning the fuel system on the composite vehicle EXCEPT:

    A. The fuel transfer pump is mechanical.

    B. The injectors are fired in pairs.

    C. Injectors 1 and 2 share a common power supply.

    D. The system uses primary and secondary fuel filters.

TASK D.7

    Answer A is incorrect. The transfer pump is a mechanical fuel pump. The pressure specification at rated RPM is between 90 and 100 psi.

    Answer B is correct. The injectors are fired individually, though they do share common supply wires.

    Answer C is incorrect. Injector power supply wires are common for 1 and 2, 3 and 4, and 5 and 6.

    Answer D is incorrect. The system uses primary and secondary fuel filters. The secondary filter supplies fuel to the engine and the after treatment system.

TASK B.4

43. Refer to the composite vehicle to answer this question: Which of the following would be LEAST LIKELY to cause the instrument panel temperature gauge to show higher than normal operating coolant temperature?

   A. Restricted coolant passages in the radiator

   B. Higher-than-normal resistance in the engine coolant temperature (ECT) sensor

   C. Restricted external passages through the charge air cooler

   D. A damaged water pump impeller

Answer A is incorrect. Restricted coolant passages in the radiator could raise engine temperature and cause a higher operating temperature.

Answer B is correct. The ECT sensor is a thermister. The higher the resistance, the lower the indicated temperature will be. Higher-than-normal resistance would produce a lower-than-normal instrument panel gauge reading.

Answer C is incorrect. The charge air cooler is normally mounted in front of the radiator. If the cooler was restricted, then the airflow through the radiator would be restricted, raising coolant temperature.

Answer D is incorrect. A damaged water pump impeller will reduce coolant flow and will raise coolant temperature.

TASK B.5

44. Refer to the composite vehicle (diagram on page 38 of the booklet) to answer this question: There is an active DTC set for exhaust backpressure out-of-range high. Any of the following could be the cause EXCEPT:

   A. A restricted DOC.

   B. A short between terminals 1 and 3 in connector AX.

   C. A short between terminals 1 and 2 in connector AX.

   D. A restricted DPF.

Answer A is incorrect. A restriction in the DOC will cause exhaust to back up in the exhaust system and could set an active DTC for high exhaust pressure.

Answer B is incorrect. A short between terminals 1 and 3 in connector AX would cause high voltage to be sent to the ECM on circuit 301. The ECM would interpret this as an out-of-range high fault.

Answer C is correct. A short between terminals 1 and 2 in connector AX would cause low voltage to be sent to the ECM on circuit 301. The ECM would interpret this as an out-of-range low fault.

Answer D is incorrect. A restriction in the DPF will cause exhaust to back up in the exhaust system and could set an active DTC for high exhaust pressure.

45. Refer to the composite vehicle to answer this question: The composite vehicle is hard to start after sitting overnight. What should the technician do to isolate the problem?

TASK D.2

　　A.　Advise the driver to spray ether into the intake manifold to help start the engine.

　　B.　Use the hand primer pump to determine if the fuel system has lost prime.

　　C.　Remove the injectors and bench test them.

　　D.　Replace the ECM cooling plate.

Answer A is incorrect. The technician needs to determine the cause of the hard to start concern. While spraying ether into the intake manifold may help the engine start, it does not fix the problem and is not something that should be recommended to the driver.

Answer B is correct. If the technician finds little or no resistance on the hand primer pump, there is reason to believe the fuel system has lost prime.

Answer C is incorrect. The fuel system needs to be checked for air-in-the-fuel problems before removing and testing injectors.

Answer D is incorrect. The ECM cooling plate may be a source of air in the fuel. However, first the technician needs to determine whether air in the fuel is the concern and then isolate the different areas of the fuel system to locate the air leak.

PREPARATION EXAM 5 – ANSWER KEY

1.	C	21.	C	41.	D
2.	D	22.	A	42.	C
3.	C	23.	C	43.	A
4.	C	24.	C	44.	A
5.	D	25.	A	45.	B
6.	D	26.	D		
7.	B	27.	D		
8.	A	28.	D		
9.	C	29.	B		
10.	A	30.	C		
11.	B	31.	C		
12.	A	32.	B		
13.	D	33.	B		
14.	C	34.	D		
15.	C	35.	C		
16.	A	36.	C		
17.	A	37.	D		
18.	B	38.	A		
19.	D	39.	A		
20.	B	40.	C		

PREPARATION EXAM 5 – EXPLANATIONS

TASK A.15

1. Technician A says a restricted air cleaner can cause slow acceleration. Technician B says a restricted diesel oxidation catalyst (DOC) can cause slow acceleration. Who is correct?

    A. A only

    B. B only

    C. Both A and B

    D. Neither A nor B

    Answer A is incorrect. Technician B is also correct.

    Answer B is incorrect. Technician A is also correct.

    Answer C is correct. Both Technicians are correct. Anything that restricts airflow through the engine in the air intake or in the exhaust can, and will, cause slow acceleration.

    Answer D is incorrect. Both Technicians are correct.

EGR temperature	70°F (21.1°C)
Coolant temperature	50°F (10°C)
Oil temperature	70°F (21.1°C)
Ambient air temperature	70°F (21.1°C)

2. Refer to the table above. The driver of a vehicle complains of poor fuel economy. After the vehicle has been sitting on the lot for 24 hours, the technician connects the scan tool and finds there are no active or inactive DTCs, then reviews the sensor values listed in the table. Technician A says the coolant is diluted and needs to be drained and refilled. Technician B says the ambient air temperature sensor is indicating an incorrect temperature and needs to be replaced. Who is correct?

    TASK B.2

    A. A only

    B. B only

    C. Both A and B

    D. Neither A nor B

    Answer A is incorrect. The engine coolant temperature (ECT) sensor is showing a substantially lower temperature than the other sensors. After sitting for 24 hours, the sensors should indicate nearly the same temperature. A shifted coolant temperature sensor is indicated; the circuit should be diagnosed.

    Answer B is incorrect. The ambient air temperature sensor is indicating a temperature in accord with the other sensors. The coolant temperature value is showing incorrectly.

    Answer C is incorrect. Neither Technician is correct.

    Answer D is correct. Neither Technician is correct.

3. Which of the following air inlet restriction test results indicate that the air filter needs to be changed?

    TASK C.1

    A. 5 in. H_2O

    B. 1 in. Hg

    C. 25 in. H_2O

    D. 30 in. Hg

    Answer A is incorrect. Five inches of water is a very low restriction and is acceptable. The filter does not need to be replaced.

    Answer B is incorrect. The air inlet restriction test is performed using the water manometer, not the mercury manometer.

    Answer C is correct. Twenty-five inches of water is considered to be the maximum allowable restriction of the air filter. When the air inlet restriction reaches 25 inches of water, the air filter needs to be replaced.

    Answer D is incorrect. The mercury scale is not used for this test. The water scale is the correct scale.

TASK B.2

4. Technician A says if an image is taken of the ECM prior to performing repairs, the Technician will have a before-and-after documentation of repairs. Technician B says downloading and saving ECM data before replacing the ECM is advisable. Who is correct?

    A.   A only

    B.   B only

    C.   Both A and B

    D.   Neither A nor B

    Answer A is incorrect. Technician B is also correct.

    Answer B is incorrect. Technician A is also correct.

    Answer C is correct. Both Technicians are correct. Saving an image will allow the technician to explain repairs to the customer easily. If data can be retrieved from a failed ECM, it will be easier for the technician to properly set up the new ECM.

    Answer D is incorrect. Both Technicians are correct.

TASK D.2

5. The turbocharger was replaced due to a worn compressor wheel. The engine is still low on power. Which of the following is the LEAST LIKELY cause?

    A.   The piston rings are worn.

    B.   The primary fuel filter is restricted.

    C.   The secondary fuel filter is restricted.

    D.   The ECM is faulty.

    Answer A is incorrect. If the turbocharger impeller blades were worn, it is possible that the engine had been subjected to dust, and the compression rings are worn out. This would cause low power.

    Answer B is incorrect. A restricted primary fuel filter is a common cause of low power.

    Answer C is incorrect. A restricted secondary fuel filter is a common cause of low power.

    Answer D is correct. A faulty ECM can cause a no-start condition, set inaccurate codes, or fail to operate actuators. A low power complaint, however, is not a failure associated with a failed ECM.

After treatment diesel oxidation catalyst (DOC) inlet temperature	70°F (21.1°C)
After treatment diesel particulate filter (DPF) inlet temperature	600°F (315.5°C)
After treatment diesel particulate filter (DPF) outlet temperature	500°F (260°C)

6. Referring to the table above, the readings shown were taken 10 minutes after a stationary regeneration was started. Which of the following is indicated?

   A.   A restricted DPF

   B.   A damaged DOC

   C.   A damaged DPF

   D.   A damaged DOC inlet temperature sensor

TASK B.5

   Answer A is incorrect. The stationary regeneration is performed to clean a restricted DPF. The DOC inlet temperature is much lower than normal. The after treatment system is not operating correctly.

   Answer B is incorrect. A damaged DOC is not indicated. The DOC inlet temperature is much lower than is realistic; either the sensor or the circuit is faulty.

   Answer C is incorrect. The DOC inlet temperature is lower than expected. A damaged DPF can be indicated by a high-pressure differential across the DPF.

   Answer D is correct. The DOC inlet temperature sensor is indicating a temperature that is far too low. Either the sensor or the circuit is faulty.

7. A truck equipped with an HPCR fuel system has an active DTC that reads, "unable to achieve desired common rail pressure." Which of the following is the LEAST LIKELY cause?

   A.   A restricted primary fuel filter

   B.   A restricted high-pressure drain line overflow

   C.   Worn injectors

   D.   A restricted secondary fuel filter

TASK D.5

   Answer A is incorrect. A restricted primary fuel filter can prevent adequate fuel flow and starve the HP pump. This will create low pressure in the common rail.

   Answer B is correct. A restricted drain line on the injectors can cause low power; however, it will not cause the pump to create insufficient pressure.

   Answer C is incorrect. Worn injectors will leak the high-pressure fuel and cause this code.

   Answer D is incorrect. A restricted secondary fuel filter would cause the HP pump to receive inadequate fuel delivery and create high pressure.

8. A technician is testing the resistance of the glow plug. Any of the following would indicate an unacceptable resistance EXCEPT:

   A.   0.5 ohms.

   B.   5.0 ohms.

   C.   50.0 ohms.

   D.   Out of limit (OL) ohms.

TASK C.5

   Answer A is correct. Typical glow plug resistance specifications are less than 1.0 ohm.

   Answer B is incorrect. A resistance reading of 5.0 ohms is greater than specification and may result in insufficient heat production.

   Answer C is incorrect. A resistance reading of 50.0 ohms is greater than specification and would result in practically no heat production.

   Answer D is incorrect. A resistance reading of OL ohms indicates an open circuit. No heat would be produced.

TASK B.10

9. A diesel engine has had multiple EGR coolers replaced. Technician A says a leaking head gasket could be the cause. Technician B says a leaking cooling system could be the cause. Who is correct?

A. A only

B. B only

C. Both A and B

D. Neither A nor B

Answer A is incorrect. Technician B is also correct.

Answer B is incorrect. Technician A is also correct.

Answer C is correct. Both Technicians are correct. Often EGR coolers are located high on the engine. This location means that when air is in the cooling system, it can get trapped in the EGR cooler. Air acts as an insulator and will cause EGR cooler overheating and failure.

Answer D is incorrect. Both Technicians are correct.

Engine RPM	275 rpm
HPCR fuel pressure	250 psi
Intake air temperature	75°F (23.9°C)

TASK D.5

10. Referring to the table above, these scan tool readings were taken while cranking an engine equipped with a high-pressure common rail (HPCR) fuel system that will not start. Which of the following is the LEAST LIKELY cause of the no-start condition?

A. A failed EPS

B. A restricted primary fuel filter

C. A restricted secondary fuel filter

D. A failed HPCR pump

Answer A is correct. The EPS indicates a valid cranking speed. There is no indication that the EPS has failed.

Answer B is incorrect. The fuel pressure is too low for an HPCR fuel system. A restricted primary filter could be the cause.

Answer C is incorrect. A restricted secondary fuel filter can cause low fuel pressure. The engine will not start due to insufficient fuel pressure.

Answer D is incorrect. A failed HPCR pump can produce insufficient fuel pressure for the engine to start. A failed pump could be the cause of this no-start condition.

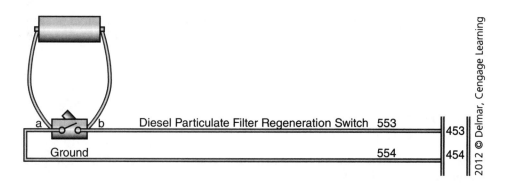

11. Referring to the figure above, when a digital volt-ohmmeter (DVOM) is connected, the DVOM meter reads 12 volts, regardless of switch position. Which of the following is the most likely cause?

TASK B.8

   A. A stuck closed (shorted) switch

   B. A stuck open (open) switch

   C. An open in circuit 553

   D. An open in circuit 554

Answer A is incorrect. A stuck closed switch will pass voltage at all times. Since this is a series circuit with no load, circuit 553 is pulled to ground when the switch is closed and the voltage drop on the meter will indicate 0.0 volts.

Answer B is correct. If the switch were stuck open, there would always be 12 volts at connector switch B and a ground at switch connector A. Therefore, the meter would indicate 12 volts at all times.

Answer C is incorrect. If circuit 553 were open, 12 volts would not be reaching the switch. The meter would indicate 0.0 volts at all times.

Answer D is incorrect. If circuit 554 were open, there would never be a ground. For the voltmeter to indicate 12 volts, there must be a ground.

12. At hot idle, the stop engine light (SEL) began flashing and then the engine shut down. Any of the following could be the cause EXCEPT:

TASK A.5

   A. The idle oil pressure reading shows 4 psi.

   B. The engine coolant temperature is 242°F (116.7°C).

   C. The engine oil temperature is 265°F (129.4°C).

   D. The coolant level is below normal level.

Answer A is correct. Idle oil pressure below 2 psi will cause a shutdown, but not at 4 psi.

Answer B is incorrect. Engine coolant temperature in excess of 240°F (115.6°C) will cause a shutdown.

Answer C is incorrect. Engine oil temperature in excess of 260°F (126.7°C) will result in a shutdown.

Answer D is incorrect. Coolant level below normal will cause a shutdown.

TASK C.3

13. Technician A says a stuck open wastegate can cause excessive intake manifold pressure. Technician B says a stuck open wastegate can cause excessive exhaust backpressure. Who is correct?

    A. A only
    B. B only
    C. Both A and B
    D. Neither A nor B

    Answer A is incorrect. A stuck open wastegate can cause low intake manifold pressure and a low power concern. It will not cause high intake manifold pressure.

    Answer B is incorrect. A stuck open wastegate can cause the engine to be slow to build intake manifold pressure, but it will not raise exhaust backpressure.

    Answer C is incorrect. Neither Technician is correct.

    Answer D is correct. Neither Technician is correct.

TASK D.5

14. Which of the following would LEAST LIKELY be used to diagnose a misfiring cylinder on a diesel engine equipped with an HPCR fuel system?

    A. A 1000 ml beaker
    B. An ohmmeter
    C. A height gauge
    D. A scan tool

    Answer A is incorrect. A 1000 ml beaker is used to measure fuel return. This test is used to find an injector that is leaking fuel.

    Answer B is incorrect. An ohmmeter can be used to measure the resistance of the injector solenoid.

    Answer C is correct. A height gauge can be used to measure the height of an EUI injector, but it is not used on an HPCR fuel system.

    Answer D is incorrect. A scan tool can be used to isolate a misfiring injector.

TASK B.4

15. Technician A says sticking swinging vanes in the VGT can cause an intermittent power loss. Technician B says a sticking EGR valve can cause intermittent power loss. Who is correct?

    A. A only
    B. B only
    C. Both A and B
    D. Neither A nor B

    Answer A is incorrect. Technician B is also correct.

    Answer B is incorrect. Technician A is also correct.

    Answer C is correct. Both Technicians are correct. If the swinging vanes stick, the VGT will not respond appropriately to commands from the ECM and turbo boost pressure can be reduced. If the EGR valve sticks open, the air coming into the engine can contain excess exhaust gas, which will lower power.

    Answer D is incorrect. Both Technicians are correct.

16. Technician A says a dial indicator can be used to check wastegate actuator travel. Technician B says the wastegate actuator arm must be removed from the turbocharger to measure its travel. Who is correct?

TASK C.3

   A. A only
   B. B only
   C. Both A and B
   D. Neither A nor B

   Answer A is correct. Only Technician A is correct. Some manufacturers recommend using a dial indicator and regulated shop air pressure to measure the travel of the wastegate actuator.

   Answer B is incorrect. To measure the travel of the wastegate actuator, the actuator must remain on the turbocharger. The distance the actuator travels is affected by the wastegate itself.

   Answer C is incorrect. Only Technician A is correct.

   Answer D is incorrect. Technician A is correct.

17. Technician A says excessive pressure in the HPCR can be caused by a faulty o-ring on the HPCR fuel system. Technician B says excessive pressure in the HPCR fuel system can be caused by a leaking high-pressure overflow valve. Who is correct?

TASK D.5

   A. A only
   B. B only
   C. Both A and B
   D. Neither A nor B

   Answer A is correct. Only Technician A is correct. The HP pump is an inlet metering style pump. If the o-ring on the control valve is damaged or missing, excessive fuel can enter the HP pump and thus cause excessive fuel pressure in the common rail.

   Answer B is incorrect. The overflow valve only opens when pressure exceeds the maximum allowable. If the valve was stuck open, the HPCR fuel system would create lower-than-specified pressure, not higher.

   Answer C is incorrect. Only Technician A is correct.

   Answer D is incorrect. Technician A is correct.

Cylinder 1	15 amps
Cylinder 2	0 amp
Cylinder 3	15 amps
Cylinder 4	0 amp
Cylinder 5	15 amps
Cylinder 6	0 amp
Cylinder 7	15 amps
Cylinder 8	0 amp

TASK C.5

18. Referring to the results from diagnostic measurements in the table above, a diesel engine equipped with glow plugs will not start in cold weather. The scan tool and an amp clamp are used to measure individual glow plug amperage. Which of the following is the most likely cause of the hard to start condition?

   A.   An open glow plug timer

   B.   An open glow plug harness

   C.   Low compression

   D.   A faulty engine position sensor

   Answer A is incorrect. An open glow plug timer will prevent all of the glow plugs from operating.

   Answer B is correct. Glow plugs for cylinders 2, 4, 6, and 8 have no current draw; it is probable that the wiring harness to that bank is open and should be repaired.

   Answer C is incorrect. Low compression can cause an engine to be hard to start in cold weather; however, the test results indicate a faulty glow plug harness.

   Answer D is incorrect. A faulty EPS can cause a diesel engine to fail to start; however, the test results indicate a failed glow plug circuit.

TASK B.2

19. Refer to the composite vehicle to answer this question: The composite diesel engine has low power and has set an active DTC for low pressure at the after treatment fuel pressure sensor. Any of the following could be the cause EXCEPT:

   A.   A restricted primary fuel filter.

   B.   A restricted secondary fuel filter.

   C.   A restriction in the distribution block.

   D.   A restricted ECM cooling plate.

   Answer A is incorrect. A restricted primary fuel filter can cause low fuel flow/pressure in the fuel system, resulting in this DTC.

   Answer B is incorrect. A restricted secondary filter can cause low fuel pressure. This low fuel pressure could cause a DTC for the after treatment fuel pressure sensor.

   Answer C is incorrect. A restriction in the distribution block can cause low fuel flow and pressure. This can cause a low fuel pressure DTC.

   Answer D is correct. The ECM cooling plate is located after the after treatment fuel pressure sensor. A restricted ECM cooling plate would raise pressure at the sensor, not lower it.

20. Refer to the composite vehicle to answer this question: The diesel particulate filter (DPF) requires regeneration more often than normal. Which of the following would be the most likely cause?

TASKS
B.8, B.10

A. The fan control switch failed open

B. Use of the wrong diesel fuel

C. An open engine brake solenoid winding

D. An open at connector K Terminal 1

Answer A is incorrect. A fan control switch that has failed open can prevent the manual override of the fan, but would not cause the DPF to need frequent regeneration.

Answer B is correct. Using low-sulfur diesel instead of the specified ultra-low sulfur diesel (ULSD) will create more soot, causing the DPF to load sooner and need to regenerate more often.

Answer C is incorrect. An open engine brake solenoid winding would affect engine brake operation, but would not cause the DPF to need frequent regeneration.

Answer D is incorrect. An open at connector K terminal 1 would result in a low coolant level sensor DTC. This would not affect how frequently DPF regeneration would be needed.

21. Refer to the composite vehicle to answer this question: The composite diesel engine has had the turbocharger replaced due to a separated turbine wheel. Now the engine lacks power. Technician A says the DOC may be restricted. Technician B says the electronic control module (ECM) should be checked for diagnostic trouble codes (DTCs). Who is correct?

TASK A.17

A. A only

B. B only

C. Both A and B

D. Neither A nor B

Answer A is incorrect. Technician B is also correct.

Answer B is incorrect. Technician A is also correct.

Answer C is correct. Both Technicians are correct. A damaged turbocharger can dump oil into the exhaust stream and cause face plugging of the DOC. When a vehicle is driven after repairs, but is still indicating symptoms of poor performance, the technician should check for any DTCs.

Answer D is incorrect. Both Technicians are correct.

2012 © Delmar, Cengage Learning

TASK B.6

22. Refer to the composite vehicle to answer this question: During a stationary regeneration, the lamp above is illuminated. Technician A says this indicates a high exhaust temperature on EGT3. Technician B says this indicates an overheat condition, and the stationary regeneration should be stopped. Who is correct?

A. A only

B. B only

C. Both A and B

D. Neither A nor B

Answer A is correct. Only Technician A is correct. This lamp indicates a high exhaust temperature on EGT3, which is a normal condition during a stationary regeneration.

Answer B is incorrect. High exhaust temperature on EGT3, as indicated by the illuminated lamp, is a normal condition for a stationary regeneration.

Answer C is incorrect. Only Technician A is correct.

Answer D is incorrect. Technician A is correct.

TASK B.3

23. Refer to the composite engine to answer this question: Technician A says that tire revolutions per mile is an adjustable parameter in the ECM. Technician B says the idle shutdown timer is an adjustable parameter in the ECM. Who is correct?

A. A only

B. B only

C. Both A and B

D. Neither A nor B

Answer A is incorrect. Technician B is also correct.

Answer B is incorrect. Technician A is also correct.

Answer C is correct. Both Technicians are correct. Tire revolutions per mile, idle shutdown timer, final drive ratio, and accelerator upper droop are some of the programmable parameters on the composite vehicle.

Answer D is incorrect. Both Technicians are correct.

Cylinder 1	0 rpm
Cylinder 2	0 rpm
Cylinder 3	100 rpm
Cylinder 4	100 rpm
Cylinder 5	100 rpm
Cylinder 6	100 rpm

24. Refer to the composite vehicle to answer this question: The results from a power balance test on the composite engine are shown in the table above. Which of the following is the most likely cause for these readings?

TASK D.1

   A. An open solenoid on injector #1

   B. A leaking charge air cooler gasket

   C. An open circuit at splice S11

   D. An open circuit at splice S13

   Answer A is incorrect. An open solenoid on injector #1 would explain the results for cylinder 1, but would not affect cylinder 2.

   Answer B is incorrect. A leaking charge air cooler will cause a boost pressure loss; however, it will not cause two cylinders to have low power.

   Answer C is correct. An open circuit at splice S11 would cause low test results on cylinders 1 and 2.

   Answer D is incorrect. An open at splice S13 would cause low test results on cylinders 5 and 6.

25. Refer to the composite vehicle (diagram on page 39 of the booklet) to answer this question: Technician A says circuits 147 and 148 are a twisted pair. Technician B says the vehicle speed sensor (VSS) is connected directly to the J1939 data bus. Who is correct?

TASK B.9

   A. A only

   B. B only

   C. Both A and B

   D. Neither A nor B

   Answer A is correct. Only Technician A is correct. The composite vehicle wiring diagram indicates that this is a twisted pair. This is to reduce electromagnetic interference.

   Answer B is incorrect. The VSS information is broadcasted across the J1939 data bus. The sensor is not directly connected to the data bus, however; instead, it is directly connected to the ECM and the ECM broadcasts the information across the data bus. The variable geometry turbine (VGT) actuator is an example of a smart device that can broadcast directly on the J1939 data bus.

   Answer C is incorrect. Only Technician A is correct.

   Answer D is incorrect. Technician A is correct.

TASK B.4

26. Refer to the composite vehicle to answer this question: Which of the following would LEAST LIKELY be used to locate a misfiring cylinder on the composite engine?

   A. An ohmmeter

   B. A scan tool

   C. A pyrometer

   D. A fuel pressure gauge

   Answer A is incorrect. An ohmmeter can be used to measure the resistance of the solenoid.

   Answer B is incorrect. A scan tool can be used to isolate a misfiring injector.

   Answer C is incorrect. A pyrometer can be used to measure exhaust temperature on individual cylinders to help isolate a misfiring cylinder.

   Answer D is correct. There is no valid test that can be performed using a fuel pressure gauge on an electronic unit injector (EUI) fuel system to help isolate a faulty injector.

TASK B.2

27. Refer to the composite vehicle to answer this question: The composite diesel engine will not start. The technician connects the scan tool and finds active DTCs for the following sensors: intake manifold temperature (IMT), intake manifold pressure, EGT1, and EPS1. Which of the following is the most likely cause of the no-start?

   A. Intake manifold temperature

   B. Intake manifold pressure

   C. EGT1

   D. Engine position sensor 1

   Answer A is incorrect. An intake manifold temperature sensor can cause the fan to operate continuously, but will not cause the engine to fail to start.

   Answer B is incorrect. A failed intake manifold pressure sensor will affect fuel delivery and can cause low power, but will not cause a no-start.

   Answer C is incorrect. An EGT1 sensor failure can affect exhaust gas recirculation (EGR) operation and DPF regeneration, but will not cause a no-start.

   Answer D is correct. A failed engine position sensor 1 will not send a correct signal to the ECM. If the ECM does not have accurate signals from both engine position sensors, it cannot determine correct engine position and the engine will fail to start.

TASKS B.10, D.3

28. Refer to the composite vehicle to answer this question: Technician A says fuel pressure should stay constant regardless of engine RPM. Technician B says excessive fuel pressure can be caused by a stuck open pressure regulator valve. Who is correct?

   A. A only

   B. B only

   C. Both A and B

   D. Neither A nor B

   Answer A is incorrect. The fuel system pressure specification is 20 psi at cranking and 90–100 psi at rated speed.

   Answer B is incorrect. A stuck open pressure regulator valve will result in unrestricted return fuel flow; this will cause low fuel pressure, not high fuel pressure.

   Answer C is incorrect. Neither Technician is correct.

   Answer D is correct. Neither Technician is correct.

29. Refer to the composite engine to answer this question: The truck overheats and shuts down. Which of the following could be the cause?

TASK A.6

    A. A shorted cooling fan control switch

    B. Circuit 158 shorted to battery positive

    C. An open engine cooling fan solenoid winding

    D. A leaking air line to the fan clutch

Answer A is incorrect. A shorted cooling fan switch would cause the fan to stay engaged all the time. This would not cause the engine to overheat and shut down.

Answer B is correct. If circuit 158 was shorted to battery positive, the solenoid would stay energized all the time. This would cause the fan to stay disengaged all the time, which would cause the truck to overheat and shut down.

Answer C is incorrect. An open engine cooling fan solenoid winding will prevent the solenoid from energizing; therefore, the fan will run continuously.

Answer D is incorrect. A leaking air line to the fan clutch will cause the fan to run continuously.

30. Refer to the composite vehicle (diagram on page 37 of the booklet) to answer this question: The composite diesel engine has active DTCs for the engine oil temperature (EOT) sensor and the inlet air temperature (IAT) sensor. Which of the following is the LEAST LIKELY cause?

TASK B.9

    A. An open circuit at splice S32

    B. An open circuit at splice S31

    C. An open circuit at ECM connector 1 terminal 25

    D. An open in circuit 222

Answer A is incorrect. An open circuit at splice S32 would only affect these two sensors. A ground at this point would affect all the sensors connected to circuit 222.

Answer B is incorrect. An open circuit at splice S31 could only affect these two sensors.

Answer C is correct. An open circuit at ECM connector 1 terminal 25 would only affect the IAT sensor signal.

Answer D is incorrect. An open in circuit 222 could affect the ECT, IMT, fuel temperature sensor (FTS), EOT, and IAT sensors.

31. Refer to the composite vehicle to answer this question: All of the following are true concerning the affects of DPF soot loading EXCEPT:

TASK C.6

    A. At full soot load the check engine light (CEL) is illuminated.

    B. Derate begins at full soot load.

    C. The DPF is overfull at 81 percent soot load.

    D. Power derate is 80 percent when the DPF is overfull.

Answer A is incorrect. The CEL is illuminated at full soot load. This is a level 3 load.

Answer B is incorrect. The power is derated 40 percent at full soot load. A full soot load is considered a level 3 load.

Answer C is correct. The composite vehicle guide does not indicate what percent soot load is considered overfull.

Answer D is incorrect. When the DPF is considered overfull (level 4), power derate is 80 percent.

TASK C.6

32. Refer to the composite vehicle to answer this question: The after treatment fuel pressure sensor indicates fuel pressure varying with engine RPM from 50–90 psi any time the engine is running. Which of the following is the most likely cause?

    A. The sensor is faulty.
    B. The after treatment fuel shutoff valve is open.
    C. The after treatment fuel shutoff valve is closed.
    D. The after treatment drain valve is open.

    Answer A is incorrect. Given that the indicated pressure varies with engine RPM it is very likely that there is fuel pressure present.

    Answer B is correct. If the after treatment fuel shutoff valve were open, this could allow fuel pressure to be present at the sensor at all times. There should be fuel pressure at the sensor only after start up and during the first minute of engine operation, or during an active regeneration of the DPF.

    Answer C is incorrect. If the after treatment fuel shutoff valve were closed, fuel pressure would not be allowed to reach the sensor.

    Answer D is incorrect. If the after treatment drain valve were open, fuel pressure would not build in the system.

TASK B.9

33. Refer to the composite vehicle to answer this question: The after treatment DPF inlet temperature sensor needs to be replaced. Where is the sensor located on the vehicle?

    A. At the inlet of the DOC
    B. Between the DOC and DPF
    C. At the outlet of the DPF
    D. At the outlet of the turbocharger

    Answer A is incorrect. The after treatment DOC inlet temperature sensor is located at the DOC inlet, not the after treatment DPF inlet temperature sensor. This is also referred to as EGT1.

    Answer B is correct. The after treatment DPF inlet temperature sensor is located after the DOC and before the DPF. This is also referred to as EGT2.

    Answer C is incorrect. The after treatment DPF outlet temperature sensor is located at the outlet of the DPF. This is also referred to as EGT3.

    Answer D is incorrect. Some new engines, 2010 and later, will have a NOx sensor located at the outlet of the turbocharger.

Exhaust backpressure	110 in. Hg
Inlet air temperature	100°F (37.8°C)
Coolant temperature	180°F (82.2°C)
Intake manifold pressure	50 psi

34. Refer to the composite vehicle to answer this question: The composite diesel engine has a low power concern. The scan tool data shown in the table above was captured with the vehicle under a medium load at 55 mph. Which of the following could be the cause of the low power concern?

TASK B.10

   A.   Low intake manifold pressure
   B.   High coolant temperature
   C.   High intake air temperature
   D.   High exhaust backpressure

Answer A is incorrect. Low intake manifold pressure could cause low power; however, this information indicates that the intake manifold pressure is normal.

Answer B is incorrect. High coolant temperature can cause the ECM to derate; however, the coolant temperature is not high, according to the scan tool data displayed.

Answer C is incorrect. High air inlet temperature can cause the engine to have low power; however, the air inlet temperature displayed would not be considered high.

Answer D is correct. High exhaust backpressure is a source of low power. The exhaust backpressure displayed is higher than normal. This could be the cause of the low power concern.

Battery voltage	13.6 volts
Cranking RPM	95 rpm
Coolant temperature	25°F (−3.9°C)
Intake air temperature	23°F (−5°C)

35. Refer to the composite vehicle to answer this question: The vehicle is hard to start, especially on a cold morning. The measurements recorded in the table above were taken with the engine cranking. Which of the following could be the cause of the concern?

TASK B.6

   A.   Low battery voltage
   B.   Key-off engine-off battery drain
   C.   High resistance in the battery cables
   D.   Incorrect coolant temperature sensor data

Answer A is incorrect. Battery voltage of 13.6 volts while cranking is very high. It does not indicate a battery that has low voltage; however, 9.0 volts would indicate a battery with low voltage.

Answer B is incorrect. If the vehicle had a key-off engine-off battery drain, the battery voltage would be much lower.

Answer C is correct. The battery voltage is high and cranking RPM is low. It is very possible that the battery cables have high resistance. This could be the source of the hard to start concern.

Answer D is incorrect. The scan tool does not indicate abnormal intake air temperature. An abnormal temperature in this situation would be 80°F (26.7°C).

TASK B.2

36. Refer to the composite vehicle booklet to answer this question: The composite vehicle has an inactive DTC stored for high fuel temperature. Which of the following is the LEAST LIKELY cause of the inactive code?

    A. Low fuel level in the tank
    B. An intermittent short at connector H
    C. High resistance at ECM connector 1 terminal 20
    D. High fuel system temperature occurred

    Answer A is incorrect. The fuel system is constantly returning fuel to the tank to help cool the fuel system components. A low fuel level in the tank can cause the overheating of the fuel; this would most likely occur on a hot day.

    Answer B is incorrect. An intermittent short at connector "H' could cause circuit resistance to be lower than normal; this would be registered in the ECM as high fuel temperature.

    Answer C is correct. High resistance at ECM connector 1 terminal 20 would cause the ECM to register the intake manifold temperature as lower than normal.

    Answer D is incorrect. If the fuel system temperature were above normal, the ECM would set a DTC for high fuel system temperature. When the fuel cooled, the active code would become inactive.

TASK B.10

37. Refer to the composite vehicle to answer this question: The composite engine has an active DTC for the fuel temp sensor. The technician checks voltage at connector H. With the connector connected to the sensor, the voltmeter reads 0.0 volts. With the connector disconnected from the sensor, the voltmeter reads 5.0 volts. When the same test is performed at ECM connector 1 Pins 21 and 22, the voltmeter reads 5.0 volts with the sensor connected or disconnected. Technician A says the ECM has failed and must be replaced. Technician B says the sensor has failed and must be replaced. Who is correct?

    A. A only
    B. B only
    C. Both A and B
    D. Neither A nor B

    Answer A is incorrect. Technician A is incorrect.

    Answer B is incorrect. Technician B is incorrect.

    Answer C is incorrect. Neither Technician is correct.

    Answer D is correct. Neither Technician is correct. There is high resistance in the wiring harness between the ECM and the sensor. When the sensor is connected and completes the circuit the voltage drops because of the high resistance. The wiring harness needs to be repaired.

38. Refer to the composite vehicle (diagram on page 40 of the booklet) to answer this question: An ohmmeter is connected at ECM connector 4, terminals 464 and 465. With the key off, the ohmmeter reads out of limit (OL or infinity) ohms. All of the following could be the cause EXCEPT:

TASK B.9

A. An open in one of the J1939 backbone resistors.
B. An open in circuit 566 between the ECM and connector XX.
C. An open in circuit 567 between the ECM and connector XX.
D. Connector XX is unplugged.

Answer A is correct. An open in one of the backbone resistors would cause the ohmmeter to read 120 ohms of resistance because the open would be in the parallel portion of the circuit.

Answer B is incorrect. An open in circuit 566 would cause the meter to read OL because it would create an open in the series portion of the circuit.

Answer C is incorrect. An open in circuit 567 would cause the meter to read OL because it would create an open in the series portion of the circuit.

Answer D is incorrect. If connector XX were unplugged, the ohmmeter would read OL because the open would be in the series portion of the circuit.

39. Refer to the composite vehicle to answer this question: The engine runs poorly, has an active DPF fault, and blows black smoke. Technician A says worn injectors could be the cause. Technician B says a worn transfer pump could be the cause. Who is correct?

TASK D.6

A. A only
B. B only
C. Both A and B
D. Neither A nor B

Answer A is correct. Only Technician A is correct. Worn injectors can cause poor injector spray patterns and black smoke. The excessive fuel in the exhaust from worn injectors can damage the DPF.

Answer B is incorrect. A worn transfer pump can cause low fuel pressure and low power; however, it will not cause black smoke.

Answer C is incorrect. Only Technician A is correct.

Answer D is incorrect. Technician A is correct.

40. Refer to the composite vehicle to answer this question: The ECM shows a DTC for a level 3 soot load and the technician is preparing the vehicle to perform a stationary regeneration. Which of the following would be the maximum time the regeneration should take?

TASK B.7

A. 0.5 hours
B. 1.0 hours
C. 1.5 hours
D. 2.0 hours

Answer A is incorrect. The DPF may complete regeneration within 0.5 hours. That would be very quick, however, and is not the maximum amount of time allowed by the ECM.

Answer B is incorrect. One hour is a common stationary regeneration time frame; however, it is not the maximum time allowed by the ECM.

Answer C is correct. According to the composite vehicle reference booklet, the stationary regeneration of the DPF may take up to 1.5 hours.

Answer D is incorrect. The composite vehicle reference guide states that the stationary regeneration may take up to 1.5 hours. Two hours is more than the amount of time specified.

TASK B.9

41. Refer to the composite vehicle to answer this question: The engine has an active DTC for the DPF differential pressure sensor (DPF Delta P). All of the following are true concerning the sensor EXCEPT:

    A. The sensor measures pressure drop.
    B. The sensor is mounted on the DPF.
    C. The sensor has two ports.
    D. One side of the sensor monitors pressure before the DOC and the other side of the sensor monitors pressure after the DPF.

    Answer A is incorrect. The differential pressure sensor measures the pressure drop across the DPF to determine whether there is a restriction.

    Answer B is incorrect. The differential pressure sensor is mounted on the DPF.

    Answer C is incorrect. The differential pressure sensor has two ports that are connected to either side of the DPF.

    Answer D is correct. The differential pressure sensor connects to the inlet and the outlet of the DPF. Both of these connections are after the DOC. The restriction of the DOC is not included in the calculations.

TASK B.6

42. Refer to the composite vehicle (diagram on page 39 of the booklet) to answer this question: The composite vehicle has an open circuit between battery positive and S54. Which of the following is LEAST LIKELY to occur?

    A. The engine will not start.
    B. The engine ECM will not communicate with the scan tool.
    C. The engine will idle poorly.
    D. The engine ECM will not communicate with other modules on J1939.

    Answer A is incorrect. The engine will not start because the ECM is not powered.

    Answer B is incorrect. If the ECM is not powered, then it cannot communicate with the scan tool or any other device on the network.

    Answer C is correct. The engine will not start; therefore, the engine will not idle at all.

    Answer D is incorrect. The engine ECM will not communicate with any device on J1939 because it is not powered.

TASK B.2

43. Refer to the composite vehicle to answer this question: The composite diesel engine has a DTC for "invalid EPS2 sensor signal." Technician A says this can cause the engine to fail to start. Technician B says this can cause the engine to shut down. Who is correct?

    A. A only
    B. B only
    C. Both A and B
    D. Neither A nor B

    Answer A is correct. Only Technician A is correct. If the EPS2 signal is lost, the engine will fail to start. The ECM requires a valid signal from EPS1 and EPS2 to start.

    Answer B is incorrect. If EPS2 fails while the engine is running, the ECM will continue to run the engine using EPS1. However, if the engine is shut off, it will not restart with a failed EPS2.

    Answer C is incorrect. Only Technician A is correct.

    Answer D is incorrect. Technician A is correct.

44. Refer to the composite vehicle (diagram on page 37 of the booklet) to answer this question: All of the following can cause a no-start condition on the composite diesel engine EXCEPT:

TASK B.2

A. An open connector C.

B. An open connector A.

C. Failed engine position sensor 1 (EPS1).

D. Failed EPS2.

Answer A is correct. Connector C is the intake manifold pressure sensor. The engine will start with this disconnected, but it will have low power and set a DTC.

Answer B is incorrect. Connector A is for the injectors and engine brake solenoids. If this is disconnected, it will cause a no-start condition.

Answer C is incorrect. A failed EPS1 or EPS2 will prevent the engine from starting.

Answer D is incorrect. If neither engine position sensor sends a signal to the engine ECM, the ECM will not energize the injectors.

45. Refer to the composite vehicle to answer this question: The composite diesel engine runs poorly. The technician disconnects connector A and the engine continues to run. Which of the following could be the cause?

TASK B.1

A. A failed ECM

B. Failed injectors

C. A worn engine camshaft

D. A damaged tone wheel for EPS1

Answer A is incorrect. If the engine will run with the ECM disconnected from the injectors, the injectors have failed, not the ECM.

Answer B is correct. When connector A is disconnected from the ECM, the injectors are no longer being energized. If the engine continues to run, the injectors are injecting fuel without the injector solenoids being energized. Thus, the injectors are faulty.

Answer C is incorrect. A worn engine camshaft can prevent the injector from injecting fuel. It will not cause the injector to inject fuel when disconnected from the ECM.

Answer D is incorrect. A damaged tone wheel could cause a misfire or a no-start condition. It could not cause the engine to run with the injectors disconnected.

PREPARATION EXAM 6 – ANSWER KEY

1.	A	21.	A	41.	D
2.	C	22.	C	42.	C
3.	C	23.	C	43.	B
4.	A	24.	A	44.	A
5.	A	25.	D	45.	D
6.	C	26.	B		
7.	D	27.	D		
8.	C	28.	C		
9.	C	29.	A		
10.	C	30.	A		
11.	A	31.	B		
12.	C	32.	A		
13.	D	33.	A		
14.	D	34.	C		
15.	C	35.	D		
16.	A	36.	D		
17.	C	37.	D		
18.	C	38.	C		
19.	D	39.	C		
20.	A	40.	B		

PREPARATION EXAM 6 – EXPLANATIONS

Engine RPM	275 rpm
HPCR fuel pressure	250 psi
Intake air temperature	75°F (23.9°C)

TASK D.5

1. Referring to the table of readings above, an engine equipped with a high-pressure common rail (HPCR) fuel system will not start. The scan tool readings were taken while cranking the engine. Which of the following is the LEAST LIKELY cause of the no-start condition?

  A. A failed EPS
  B. Leaking injectors
  C. A leaking pressure relief valve
  D. A failed HPCR pump

  Answer A is correct. The EPS indicates a valid cranking speed. There is no indication that the EPS has failed.

  Answer B is incorrect. The fuel pressure is too low for an HPCR fuel system. HPCR fuel injectors will leak internally when worn and cause low fuel pressure.

  Answer C is incorrect. A leaking pressure relief valve will cause low fuel pressure. The engine will not start due to insufficient fuel pressure.

  Answer D is incorrect. A failed HPCR pump can produce insufficient fuel pressure for the engine to start. A failed pump could be the cause of this no-start condition.

2. The head gasket has been replaced twice on a vehicle in fewer than 5,000 miles. Technician A says the head bolts may be stretched. Technician B says incorrect injector calibration codes installed in the ECM could be the cause. Who is correct?

TASK B.1

    A. A only

    B. B only

    C. Both A and B

    D. Neither A nor B

Answer A is incorrect. Technician B is also correct.

Answer B is incorrect. Technician A is also correct.

Answer C is correct. Both Technicians are correct. Stretched head bolts lose their clamping force and can cause repeat head gasket failures. Incorrect calibration codes installed in the ECM can cause the engine to create more horsepower than it was designed to. This can cause a leaking head gasket.

Answer D is incorrect. Both Technicians are correct.

3. A vehicle with an electronic unit injector (EUI) fuel system has had all six injectors replaced due to a black smoke concern. Now the truck idles poorly. Technician A says the injector height may not have been set properly during the installation. Technician B says the injector calibration codes may not have been installed properly. Who is correct?

TASK B.11

    A. A only

    B. B only

    C. Both A and B

    D. Neither A nor B

Answer A is incorrect. Technician B is also correct.

Answer B is incorrect. Technician A is also correct.

Answer C is correct. Both Technicians are correct. Failure to set the injectors at the correct height can cause unequal fuel delivery and rough idle. Failure to install the correct injector calibration codes can cause unequal fuel delivery and rough idle.

Answer D is incorrect. Both Technicians are correct.

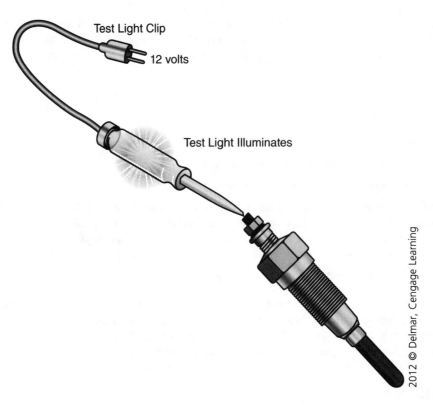

Test Light Clip

12 volts

Test Light Illuminates

2012 © Delmar, Cengage Learning

TASK D.6

4. Referring to the figure above, a diesel engine equipped with a hydraulically actuated electronically controlled unit injector (HEUI) fuel system will not start. The technician performs the test above on all the glow plugs and the test light illuminates on every glow plug. Any of the following could be the cause of the no-start condition EXCEPT:

A.   Faulty glow plugs.

B.   Low oil level in the oil pan.

C.   A faulty glow plug controller.

D.   Low injection actuation pressure.

Answer A is correct. The glow plugs passed the test. They are not faulty.

Answer B is incorrect. Low oil level can prevent HEUI injection operation. This could be the cause for the no-start condition.

Answer C is incorrect. A faulty glow plug controller can cause the glow plugs to fail to heat, resulting in a no-start condition.

Answer D is incorrect. Low injection actuation pressure can prevent HEUI injector operation. This would cause a no-start condition.

5. A vehicle that has had a turbocharger replacement now has low power and high exhaust temperature. Technician A says a portion of the old exhaust turbine may be lodged in the exhaust pipe. Technician B says the wastegate may be stuck open on the new turbocharger. Who is correct?

TASK B.11

   A. A only
   B. B only
   C. Both A and B
   D. Neither A nor B

   Answer A is correct. Only Technician A is correct. If the exhaust turbine separated from the shaft on the old turbocharger, a piece of the exhaust turbine may have lodged in the exhaust system, although the technician did not see it when replacing the turbo. This would cause exhaust restriction, high exhaust system pressure and temperature, and low power.

   Answer B is incorrect. If the wastegate were stuck open, the engine would be slow to accelerate under load, but would not have high exhaust temperature.

   Answer C is incorrect. Only Technician A is correct.

   Answer D is incorrect. Technician A is correct.

6. Technician A says a smoke machine can be used to test the integrity of the exhaust system. Technician B says a smoke machine can be used to check the integrity of the intake system. Who is correct?

TASKS A.8, A.9

   A. A only
   B. B only
   C. Both A and B
   D. Neither A nor B

   Answer A is incorrect. Technician B is also correct.

   Answer B is incorrect. Technician A is also correct.

   Answer C is correct. Both Technicians are correct. A smoke machine generates smoke to help the technician locate leaks in any enclosed system. It can be used in the intake, exhaust, or crankcase to help locate a leak.

   Answer D is incorrect. Both Technicians are correct.

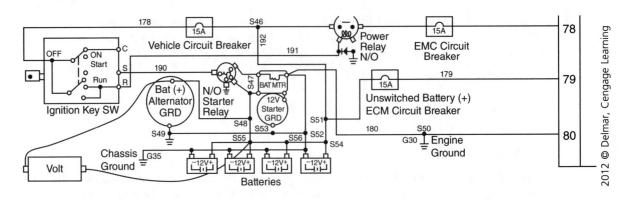

**TASKS
B.6, B.8**

7. Referring to the figure above, a voltage drop test is performed on the composite engine. With the engine running at fast idle and all the electrical accessories on the vehicle turned on, the voltmeter indicates 3.6 VDC. Which of the following could be the cause?

A. Low battery voltage

B. Excessive current draw of electrical accessories

C. A faulty alternator ground

D. High resistance between S48 and S55

Answer A is incorrect. A reading of 3.6 VDC would indicate low battery voltage. The technician is checking voltage drop, however, instead of battery voltage.

Answer B is incorrect. Excessive current draw of accessories would most likely result in blown fuses; this test is checking voltage drop.

Answer C is incorrect. Voltage drop testing is effective to check voltage drop; however, the test would need to be performed on the negative side of the circuit. The technician is checking the positive side of the circuit.

Answer D is correct. High resistance between S48 and S55 would cause excessive voltage drop. A reading of 3.6 VDC indicates excessive voltage drop; 0.5 volts would be normal.

TASK B.6

8. The diesel engine cranks slowly. Which of the following would LEAST LIKELY be used to test the batteries?

A. A load tester

B. A voltmeter

C. An ohmmeter

D. A capacitance tester

Answer A is incorrect. A load tester can be used to check the ability of the batteries to deliver the current needed to start the engine.

Answer B is incorrect. A voltmeter can be used to check the vehicle battery voltage while cranking.

Answer C is correct. An ohmmeter is not used to test the batteries; it is used to check circuit continuity.

Answer D is incorrect. A capacitance tester can be used to test the batteries; this is a newer test to see if the batteries can deliver the amperage needed to start the engine.

9. An engine has an active DTC for an accelerator pedal position (APP) sensor. The APP is a Hall-effect-style sensor. Which of the following tools would the technician most likely use to diagnose the sensor?

TASK D.4

    A.　Ohmmeter

    B.　Ammeter

    C.　Scan tool

    D.　Oscilloscope

Answer A is incorrect. An ohmmeter is an effective tool to diagnose a potentiometer-style sensor, but a Hall-effect-style sensor should be tested using a scan tool.

Answer B is incorrect. An ammeter is an effective tool to check current flow, but the Hall-effect-sensor is not being tested for current flow.

Answer C is correct. The technician needs to actuate the pedal while watching the signal on the scan tool.

Answer D is incorrect. The technician could use an oscilloscope to check the signal from the Hall-effect sensor; however, OEMs typically do not provide duty cycle or oscilloscope pattern troubleshooting information about checking the signal in this manner.

10. A diesel engine that is not equipped with a DPF cranks, but will not start. There is no smoke from the exhaust while cranking. Any of the following could be the cause EXCEPT:

TASK B.4

    A.　No voltage to the ECM.

    B.　No fuel to the engine.

    C.　Low compression.

    D.　A failed engine position sensor.

Answer A is incorrect. No voltage to the ECM would prevent fuel injection; thus, there would be no smoke and no start.

Answer B is incorrect. No fuel to the engine would mean no fuel is injected; thus, there would be no smoke and no start.

Answer C is correct. Low compression can cause the engine not to start; however, there should be smoke.

Answer D is incorrect. A failed engine position sensor will prevent fuel injections; thus, no smoke and no start.

11. A diesel engine will not start. A symbol of a key is flashing on the instrument panel. Which of the following is LEAST LIKELY to be the cause of the no-start?

TASK B.7

    A.　Low fuel level

    B.　An incorrect password entered

    C.　An incorrect key being used

    D.　A failed vehicle security system

Answer A is correct. The flashing key symbol indicates a security system issue. Low fuel level will not cause this symbol.

Answer B is incorrect. Some vehicle security systems require the driver to enter a driver ID or password. The operator could have entered the wrong password or failed to enter a password at all.

Answer C is incorrect. Some vehicle security systems use coded keys to enable and disable the security system.

Answer D is incorrect. The security system may have failed and is preventing the engine from starting.

TASK D.6

12. Which of the following would LEAST LIKELY be used to diagnose a misfiring cylinder on an HEUI fuel system?

    A. Scan tool

    B. Ohmmeter

    C. 1000 ml beaker

    D. Pyrometer

Answer A is incorrect. A scan tool can be used to run a click or buzz test on an HEUI injector.

Answer B is incorrect. An ohmmeter can be used to measure the resistance of the solenoid on an HEUI injector.

Answer C is correct. A 1000 ml beaker is used to measure the drain system flow on an HPCR fuel system. It is not used on a HEUI fuel system.

Answer D is incorrect. A pyrometer can be used to measure individual exhaust temperatures per cylinder to help diagnose a misfiring cylinder.

TASK B.11

13. The technician takes a test drive in a vehicle after replacing a failed DPF Delta P sensor. The malfunction indicator lamp (MIL) is still on. The engine is heavy-duty onboard diagnostic (HD-OBD) compliant. The SEL and the CEL are both off. The technician connects the scan tool and the engine ECM indicates no active trouble codes. Which of the following is the most likely cause?

    A. A failed ECM

    B. A failed J1939 data bus

    C. A failed scan tool

    D. Technician failed to reset the MIL

Answer A is incorrect. This can be a normal condition on an HD-OBD compliant vehicle. There is no reason to believe that the ECM has failed.

Answer B is incorrect. A failed J1939 data bus can result in no communications between the modules. It would not cause the condition described in the question.

Answer C is incorrect. The scan tool is connected and reporting correct information. The technician has not performed the necessary action.

Answer D is correct. The MIL will not be extinguished until a certain number of good trips occur, often three. Typically, the technician can extinguish this lamp through a command in the scan tool.

TASK D.5

14. An HPCR fuel system is being tested for fuel return flow from the overflow valve. Technician A says excessive fuel return can be caused by restricted fuel filters. Technician B says excessive fuel return can be caused by a restricted high-pressure relief valve. Who is correct?

    A. A only

    B. B only

    C. Both A and B

    D. Neither A nor B

Answer A is incorrect. Restricted fuel filters will cause a decreased flow, not an increased flow.

Answer B is incorrect. A restricted high-pressure valve can cause excessive pressure in the fuel rail, but not excessive flow.

Answer C is incorrect. Neither Technician is correct. Excessive flow from the high-pressure relief valve can be caused by a failed pressure regulator or a failed relief valve.

Answer D is correct. Neither Technician is correct.

15. The ECM has set a fault code for EPS1. During troubleshooting, the technician disconnects the EPS1 wiring harness and finds the internal terminals covered in engine oil. Which of the following is the most likely cause?

TASK B.9

 A. High engine oil pressure

 B. High engine crankcase pressure

 C. A failed EPS1

 D. A leaking valve cover gasket

Answer A is incorrect. The EPS1 is supposed to be sealed. Oil pressure should not be able to enter the sensor. The sensor has failed.

Answer B is incorrect. The EPS1 is a sealed sensor designed to operate in the presence of oil. If the sensor has internal oil as described in the question, the sensor has failed.

Answer C is correct. EPS1 has failed. It is no longer sealed and is allowing oil to enter.

Answer D is incorrect. A leaking valve cover gasket could allow oil to be on the outside of the sensor; however, the sensor connector is sealed and should not allow oil into the terminal area.

16. An engine with an HPCR fuel system starts okay when cold, but is hard to start when warm. The customer does not have any other concerns. Technician A says faulty injectors could be the cause. Technician B says a restricted air filter could be the cause. Who is correct?

TASK D.5

 A. A only

 B. B only

 C. Both A and B

 D. Neither A nor B

Answer A is correct. Only Technician A is correct. HPCR injectors will start to leak internally when they wear. The symptoms are usually a hot hard start concern. The injectors can be isolated using a fuel return test.

Answer B is incorrect. A restricted air filter can cause a low power concern, but would not cause a hot restart concern with no other problems.

Answer C is incorrect. Only Technician A is correct.

Answer D is incorrect. Technician A is correct.

17. Technician A says incorrect calibration files can cause poor fuel economy. Technician B says incorrect calibration files can cause poor engine performance. Who is correct?

TASK B.3

 A. A only

 B. B only

 C. Both A and B

 D. Neither A nor B

Answer A is incorrect. Technician B is also correct.

Answer B is incorrect. Technician A is also correct.

Answer C is correct. Both Technicians are correct. Incorrect calibration files can affect fuel delivery, idle quality, and power. When troubleshooting a diesel engine concern, the technician should check to make sure the correct and latest calibration file is loaded into the ECM.

Answer D is incorrect. Both Technicians are correct.

TASK D.5

18. An engine equipped with an HPCR fuel system has a DTC for "unable to reach desired fuel rail pressure." Technician A says the problem can be worn injectors. Technician B says the problem can be a worn high-pressure pump. Who is correct?

   A.   A only

   B.   B only

   C.   Both A and B

   D.   Neither A nor B

Answer A is incorrect. Technician B is also correct.

Answer B is incorrect. Technician A is also correct.

Answer C is correct. Both Technicians are correct. A worn high-pressure pump can leak fuel internally and cause the condition described. Injectors leaking internally will make it harder for the high-pressure pump to reach desired fuel rail pressure and cause this concern also.

Answer D is incorrect. Both Technicians are correct.

TASKS
A.4, B.5, C.3

19. A diesel engine that is not equipped with a DPF blows smoke and is hard to start. Any of the following could be the cause EXCEPT:

   A.   Worn compression rings.

   B.   Worn injectors.

   C.   A restricted engine air filter.

   D.   A stuck open wastegate.

Answer A is incorrect. Worn compression rings will cause low compression, smoke, and hard starting.

Answer B is incorrect. Worn injectors can deliver poor spray patterns, causing smoke and hard starting.

Answer C is incorrect. A restricted engine air filter can cause smoke, and if extremely restricted, can cause hard starting.

Answer D is correct. A stuck open wastegate would cause the engine to be slow to accelerate under load; however, it would not cause the engine to be hard to start.

TASK B.3

20. All of the following should be done prior to installing a new calibration file EXCEPT:

   A.   Warm the engine to operating temperature.

   B.   Ensure the batteries are fully charged.

   C.   Connect the scan tool to an external power supply.

   D.   Connect the scan tool to the engine ECM.

Answer A is correct. It is not necessary to warm the engine. In some cases, it may not even be possible.

Answer B is incorrect. The batteries in the vehicle should be fully charged and in good condition. Some manufacturers also recommend connecting a battery maintainer to the batteries to ensure that battery voltage does not fall during the procedure.

Answer C is incorrect. The scan tool should be connected to an external power source to ensure that the internal battery in the scan tool does not get too low during the procedure.

Answer D is incorrect. The scan tool must be connected to the ECM when installing new calibration files. Some manufacturers recommend using a dedicated data link near the engine instead of the ATA data link located in the cab.

21. Refer to the composite vehicle to answer this question: The composite vehicle has a DTC for misfire on cylinder #2. During testing, cylinder #2 fails a cylinder contribution test. A compression test is performed and compression is found to be acceptable. Any of the following could be the cause of the DTC EXCEPT:

    TASKS
    B.4, B.10

    A. A leaking exhaust valve.

    B. A worn injector cam lobe.

    C. A worn injector.

    D. High resistance in the injector solenoid.

    Answer A is correct. A leaking exhaust valve would have shown low compression when the compression test was performed.

    Answer B is incorrect. A worn injector cam lobe can cause the injector to fail to inject fuel, which would cause a misfire, but will not affect compression.

    Answer C is incorrect. A worn injector can cause a misfire because it can leak diesel fuel internally, but will not cause low compression.

    Answer D is incorrect. High resistance in the electrical solenoid can cause the injector to fail to control fuel flow and cause a misfire, but this will not cause low compression.

22. Refer to the composite vehicle to answer this question: The operator complains of the engine stalling under a long pull. Technician A says the electronic control module (ECM) should be checked for freeze frame data to help diagnose the concern. Technician B says the vehicle may need to be connected to a loaded trailer and driven to duplicate the concern. Who is correct?

    TASKS
    A.2, A.5

    A. A only

    B. B only

    C. Both A and B

    D. Neither A nor B

    Answer A is incorrect. Technician B is also correct.

    Answer B is incorrect. Technician A is also correct.

    Answer C is correct. Both Technicians are correct. If a diagnostic trouble code (DTC) has been stored that relates to this concern, the freeze frame data could be very valuable. It could indicate that the engine was shut down by the engine protection system because it was overheating. The technician will most likely need to load the vehicle to repeat the condition, either with a loaded trailer or with a dynamometer.

    Answer D is incorrect. Both Technicians are correct.

23. Refer to the composite vehicle to answer this question: Technician A says a shorted injector harness can cause a DTC. Technician B says an injector harness can be tested using an ohmmeter. Who is correct?

    TASK B.8

    A. A only

    B. B only

    C. Both A and B

    D. Neither A nor B

    Answer A is incorrect. Technician B is also correct.

    Answer B is incorrect. Technician A is also correct.

    Answer C is correct. Both Technicians are correct. A shorted injector harness will cause the injector to misfire because it will not receive the signal from the ECM. An injector wiring harness can be tested with an ohmmeter. The wires should not be continuous with each other.

    Answer D is incorrect. Both Technicians are correct.

TASKS B.2, C.10

24. Refer to the composite vehicle to answer this question: The composite diesel engine has a DTC for high crankcase pressure. Which of the following is LEAST LIKELY to be the cause?

    A. A restricted engine air filter

    B. A restricted crankcase ventilation filter

    C. Worn piston rings

    D. A hole in a piston

    Answer A is correct. A restricted engine air inlet filter can cause low power and possibly smoke, but would not cause a high crankcase pressure.

    Answer B is incorrect. A restricted crankcase ventilation filter can cause crankcase pressure to be high, resulting in a DTC.

    Answer C is incorrect. Worn piston rings allow piston ring blow-by and can cause high crankcase pressure.

    Answer D is incorrect. A hole in a piston can cause high crankcase pressure.

TASK A.10

25. Refer to the composite vehicle to answer this question: The composite engine has a knock that is evident at 800 rpm, 1200 rpm, and 2000 rpm. Which of the following is the most likely cause?

    A. A loose rod bearing

    B. A loose main bearing

    C. A loose piston pin

    D. A failed vibration damper

    Answer A is incorrect. A loose rod bearing will knock based upon load, not engine RPM.

    Answer B is incorrect. A loose main bearing is not RPM dependent; it is load dependent.

    Answer C is incorrect. A loose piston pin is not RPM dependent; a double knock is load dependent.

    Answer D is correct. Knocking noises that are evident only at certain RPMs are typically related to vibration damper concerns. Vibration dampers should be checked for leaks, radial run out, swelling, or damage.

TASK B.5

26. Refer to the composite vehicle to answer this question: The vehicle has a low power concern, but no DTCs. During diagnosis, intake manifold pressure at rated speed under full load is 25 psi. Technician A says the low power concern may be due to a missing diesel particulate filter (DPF). Technician B says the low power concern may be due to a restricted fuel filter. Who is correct?

    A. A only

    B. B only

    C. Both A and B

    D. Neither A nor B

    Answer A is incorrect. A missing DPF would not cause a low power concern.

    Answer B is correct. Only Technician B is correct. A restricted fuel filter can cause low fuel flow, which can cause low heat in the exhaust stream and low intake manifold pressure.

    Answer C is incorrect. Only Technician B is correct.

    Answer D is incorrect. Technician B is correct.

27. Refer to the composite vehicle to answer this question: The vehicle has a low power complaint. The technician verifies the concern and finds an active DTC for low intake manifold pressure. Which of the following is the LEAST LIKELY cause?

TASKS
B.4, B.7

    A. A worn turbo impeller wheel
    B. A worn turbo exhaust turbine
    C. A restricted DPF
    D. A malfunctioning theft deterrent system

    Answer A is incorrect. A worn impeller will not move enough air to create sufficient intake manifold pressure.

    Answer B is incorrect. A worn exhaust turbine will not spin the turbocharger shaft fast enough to create sufficient intake manifold pressure.

    Answer C is incorrect. A restricted DPF can cause reduced exhaust flow, leading to insufficient intake manifold pressure.

    Answer D is correct. A malfunctioning theft deterrent system can cause a no-start or starts-and-dies condition; it will not cause the engine to set a DTC for low intake manifold pressure.

28. Refer to the composite vehicle to answer this question: The customer has a low power concern. At full load, the technician finds 20 psi intake manifold pressure. There are no DTCs. Which of the following is the LEAST LIKELY cause of the concern?

TASKS
C.2, C.1

    A. Restricted fuel filters
    B. A restricted air filter
    C. An open on ECM pin 214
    D. Worn impeller blades on the turbo

    Answer A is incorrect. Restricted fuel filters can cause low fuel flow and low power; this may not set a DTC.

    Answer B is incorrect. A restricted air filter can cause reduced airflow and reduced intake manifold pressure, but would not necessarily set a DTC.

    Answer C is correct. An open on ECM pin 214 would open the private J1939 data bus; this would almost certainly set a DTC.

    Answer D is incorrect. Worn impeller blades can cause low airflow and low intake manifold pressure, but would not necessarily set a DTC.

29. Refer to the composite vehicle (diagram on pages 37 and 40 of the booklet) to answer this question: The engine cranks but fails to start. Any of the following could be the cause EXCEPT:

TASK B.6

    A. An open on circuit 554.
    B. An open circuit 223.
    C. Connector A is open.
    D. A short between ECM pins 27 and 28.

    Answer A is correct. An open on circuit 554 would prevent a regeneration from using the dash-mounted switch, but it would not cause a no-start condition.

    Answer B is incorrect. An open circuit 223 would prevent power from reaching engine position sensor (EPS) 2; this would prevent the engine from starting.

    Answer C is incorrect. If connector A were open, the injectors would not receive a signal from the ECM. This would cause a no-start condition.

    Answer D is incorrect. If ECM pins 27 and 28 were shorted, the ECM would not receive a signal from EPS2. That would prevent the engine from starting.

TASK B.6

30. Refer to the composite diesel engine (diagram on pages 37 and 39 of the booklet) to answer this question: The engine brake works on medium and high, but it does not work on low. Which of the following could be the cause?

    A. An open circuit 111
    B. An open circuit 110
    C. A failed engine brake on/off switch
    D. An open on circuit 155

    Answer A is correct. An open circuit 111 would prevent operation of the engine brake solenoids 3 and 4. This would prevent operation on low.

    Answer B is incorrect. An open circuit 110 would prevent operation on medium and make the brake weak on high.

    Answer C is incorrect. A failed brake switch would prevent the entire engine brake from working.

    Answer D is incorrect. An open on circuit 155 would prevent medium and high from working.

Engine coolant temperature	180°F (82.2°C)
Engine RPM	600 rpm
Engine oil pressure	1 psi
Exhaust backpressure	0.1 psi

TASK A.5

31. Refer to the composite vehicle to answer this question: The composite vehicle's check engine light (CEL) and stop engine light (SEL) are illuminated and the engine shuts off while idling in traffic. The freeze frame data in the table above was retrieved from the ECM. Which of the following could be the cause of the engine shutdown?

    A. The engine was overheated.
    B. The oil level was too low.
    C. The engine was idling too **fast.**
    D. There was excessive exhaust backpressure.

    Answer A is incorrect. The engine is programmed to shut down for overheating. The freeze frame data, however, does not indicate that this occurred.

    Answer B is correct. Engine oil pressure is very low; the engine protection system shut the engine off due to low oil pressure. A low oil level can cause low oil pressure.

    Answer C is incorrect. Engine RPM indicates 600 rpm. That would be a normal idle speed.

    Answer D is incorrect. Excessive exhaust backpressure can cause an engine to shut down because the exhaust gases cannot exit the engine. There is no indication of excessive exhaust backpressure in the freeze frame data, however.

32. Refer to the composite vehicle to answer this question: When the engine brake is engaged, the engine cooling fan does not engage. There are no trouble codes, and the driver has no other complaints. Technician A says incorrect programming could be the cause. Technician B says an open to the engine cooling fan solenoid could be the cause. Who is correct?

TASK B.3

    A. A only

    B. B only

    C. Both A and B

    D. Neither A nor B

    Answer A is correct. Only Technician A is correct. Whether the fan engages when the engine brake is activated is a programmable parameter. Since there are no other complaints, incorrect setting of the programmable parameter is a likely cause.

    Answer B is incorrect. An open engine cooling fan solenoid would cause the fan to operate continuously.

    Answer C is incorrect. Only Technician A is correct.

    Answer D is incorrect. Technician A is correct.

33. Refer to the composite vehicle to answer this question: Which of the following would LEAST LIKELY cause a diesel engine to start and die?

TASK B.4

    A. An open in circuit 128

    B. Air in the fuel system

    C. A loose primary fuel filter

    D. Restriction in the fuel suction line

    Answer A is correct. Circuit 128 is the signal circuit from EPS2. The ECM must see this signal to start the engine.

    Answer B is incorrect. Air in the fuel system is a common cause of a starts-and-dies complaint. The air pockets reach the fuel injectors and no fuel is delivered.

    Answer C is incorrect. A loose primary fuel filter will allow air in the fuel and cause a start-and-die situation.

    Answer D is incorrect. A restriction in the fuel suction line starves the engine for fuel, possibly causing a starts-and-dies complaint.

34. Refer to the composite vehicle to answer this question: Technician A says a restricted DPF could be caused by a restricted after treatment fuel shutoff valve. Technician B says a restricted DPF could be caused by an after treatment drain valve that has failed open. Who is correct?

TASK C.6

    A. A only

    B. B only

    C. Both A and B

    D. Neither A nor B

    Answer A is incorrect. Technician B is also correct.

    Answer B is incorrect. Technician A is also correct.

    Answer C is correct. Both Technicians are correct. A restricted after treatment fuel shutoff valve would reduce the flow of fuel to the after treatment injector; this may prevent effective regeneration of the DPF. An after treatment drain valve that has stuck open can prevent fuel pressure buildup in the after treatment fuel supply system; this would decrease fuel flow through the after treatment injector and prevent proper regeneration of the DPF.

    Answer D is incorrect. Both Technicians are correct.

TASK B.5

35. Refer to the composite vehicle to answer this question: The ECM has a DTC for high crankcase pressure. Any of the following could be the cause EXCEPT:

A. A damaged piston.

B. Worn compression rings.

C. A restricted crankcase ventilation tube.

D. A restricted engine oil filter.

Answer A is incorrect. Damaged piston rings will allow combustion chamber gases to build in the crankcase, causing high crankcase pressure.

Answer B is incorrect. Worn compression rings will cause combustion gases to enter the crankcase, causing high crankcase pressure.

Answer C is incorrect. A restricted crankcase ventilation tube can cause the pressure to build in the crankcase, resulting in high crankcase pressure.

Answer D is correct. When the engine oil filter is restricted, the oil filter pressure differential valve (by-pass valve) will open and allow unfiltered oil to circulate through the filter. This will not affect crankcase pressure.

TASK B.6

36. Refer to the composite vehicle (diagram on page 39 of the booklet) to answer this question: Which of the following would be the most likely cause of a 15 amp vehicle circuit breaker that has failed open?

A. An open on circuit 178

B. A short to ground on circuit 192

C. An open on circuit 192

D. A short to ground on circuit 178

Answer A is incorrect. An open on circuit 178 would cause no current flow. Excessive current flow is what causes a circuit breaker to fail open.

Answer B is incorrect. A short on circuit 192 would cause excessive current flow; however, it would be before the circuit breaker. Therefore, it would not affect the circuit breaker.

Answer C is incorrect. An open on circuit 192 would stop current flow. The circuit breaker would not receive any voltage.

Answer D is correct. A short to ground on circuit 178 would cause increased current flow and would cause the circuit breaker to fail open.

37. Refer to the composite vehicle (diagram on page 37 of the booklet) to answer this question: The composite engine has two active DTCs, one for the intake manifold temperature sensor and one for the intake manifold pressure sensor. There are no other active or inactive codes. Technician A says an open on circuit 222 could be the cause. Technician B says an open at splice S25 could be the cause. Who is correct?

TASK C.11

   A. A only

   B. B only

   C. Both A and B

   D. Neither A nor B

Answer A is incorrect. An open at circuit 22 would not affect only those two sensors. It may affect all the sensors on the circuit or only the sensors after the open. The way the circuit is drawn, however, it would not be possible to only affect those two sensors.

Answer B is incorrect. An open at splice S25 would affect the engine coolant temperature (ECT) sensor, intake manifold temperature sensor, fuel temperature sensor, engine oil temperature sensor, and inlet air temperature sensor. It would not affect the intake manifold pressure sensor.

Answer C is incorrect. Neither Technician is correct.

Answer D is correct. Neither Technician is correct.

38. Refer to the composite vehicle to answer this question: Technician A says global maximum road speed is a programmable parameter. Technician B says engine idle speed is a programmable parameter. Who is correct?

TASK B.3

   A. A only

   B. B only

   C. Both A and B

   D. Neither A nor B

Answer A is incorrect. Technician B is also correct.

Answer B is incorrect. Technician A is also correct.

Answer C is correct. Both Technicians are correct. Idle speed can be programmed between 600 and 850 rpm, and global maximum road speed can be programmed between 0 and 120 mph.

Answer D is incorrect. Both Technicians are correct.

39. Refer to the composite vehicle to answer this question: Technician A says a failed neutral safety switch may prevent a stationary regeneration of the DPF. Technician B says a failed accelerator can prevent a stationary regeneration of the DPF. Who is correct?

TASK B.1

   A. A only

   B. B only

   C. Both A and B

   D. Neither A nor B

Answer A is incorrect. Technician B is also correct.

Answer B is incorrect. Technician A is also correct.

Answer C is correct. Both Technicians are correct. For a stationary regeneration of the DPF to occur, there must be no input from the accelerator, brake, and clutch pedals, as well as a signal from the neutral safety switch indicating the vehicle is in neutral.

Answer D is incorrect. Both Technicians are correct.

TASK C.9

40. Refer to the composite vehicle to answer this question: All of the following are true concerning the air intake and exhaust system on the composite vehicle EXCEPT:

A. The engine has a variable geometry turbocharger (VGT).

B. The engine has a variable valve actuator (VVA).

C. The engine has cooled exhaust gas recirculation (EGR).

D. The engine has a DPF.

Answer A is incorrect. The engine is equipped with a VGT, which is a smart device that communicates over the J1939 data bus.

Answer B is correct. The composite engine does not have a VVA, although it does have cooled EGR. It would be unusual for an engine to have both.

Answer C is incorrect. The engine uses cooled EGR to help lower emissions. Cooled EGR helps lower NOx emissions.

Answer D is incorrect. The engine is equipped with a diesel oxidation catalyst (DOC) and a DPF to help lower particulate emissions.

TASK B.6

41. Refer to the composite vehicle to answer this question: The driver complains of poor performance from the air conditioner (A/C). The technician finds the engine cooling fan does not engage when the A/C is on. The fan will engage when commanded to from the scan tool. Which of the following is the most likely cause?

A. A faulty engine cooling fan solenoid

B. A failed engine cooling fan

C. A failed blower motor switch

D. A failed A/C high-side pressure switch

Answer A is incorrect. If the engine cooling fan solenoid was faulty, the fan would not operate when commanded by the scan tool.

Answer B is incorrect. If the engine cooling fan were faulty, the fan would not operate when commanded using the scan tool.

Answer C is incorrect. The blower motor switch does not control the fan; the high-side pressure switch does.

Answer D is correct. If the high-side pressure switch does not signal the ECM that high-side pressure is high, the ECM will not turn on the fan.

TASK D.8

42. Refer to the composite vehicle to answer this question: The composite engine starts and runs but has low power. There are no active or inactive DTCs. Which of the following could be the cause?

A. A missing signal from EPS1

B. A missing signal from EPS2

C. A broken injector return spring

D. An open injector solenoid winding

Answer A is incorrect. A missing signal from EPS1 would set a DTC and cause a no-start condition.

Answer B is incorrect. A missing signal from EPS2 would set a DTC and cause a no-start condition.

Answer C is correct. A broken injector return spring can cause low fuel delivery from that injector. Since the injector is still delivering a partial fuel charge, it may not set a DTC.

Answer D is incorrect. An open injector solenoid winding can cause low power; however, it would set a DTC.

43. Refer to the composite vehicle (diagram on page 39 of the booklet) to answer this question: The engine speed will increase appropriately when the remote power take-off (PTO) switch is operated; however, it will not operate when the dash-mounted PTO switch is operated. Which of the following could be the cause?

TASK B.7

   A.   An open circuit 170
   B.   An open circuit 171
   C.   An open at ECM pin 77
   D.   An open at ECM pin 70

   Answer A is incorrect. An open circuit 170 would prevent operation of the remote PTO switch.

   Answer B is correct. An open circuit 171 would prevent the signal from the PTO switch from reaching the ECM.

   Answer C is incorrect. An open at ECM pin 77 would affect all the circuits grounded on that pin.

   Answer D is incorrect. An open at ECM pin 70 would affect the remote PTO switch, not the dash-mounted switch.

44. Refer to the composite vehicle to answer this question: The composite diesel engine has a DTC for high crankcase pressure. Which of the following could be the cause?

TASK C.10

   A.   A restricted crankcase ventilation filter
   B.   Excessive engine oil pressure
   C.   A missing crankcase ventilation filter
   D.   An open on circuit 304

   Answer A is correct. A restricted crankcase ventilation filter can cause a DTC to be set for high crankcase pressure.

   Answer B is incorrect. Excessive engine oil pressure may set a DTC. The DTC would indicate excessive oil pressure but not excessive crankcase pressure.

   Answer C is incorrect. A missing crankcase filter can set a DTC. It would set a DTC for low pressure/missing filter but not for high crankcase pressure.

   Answer D is incorrect. An open on circuit 304 would cause a low signal, not a high signal. Additionally, depending upon where the circuit was open, it could cause a DTC for the exhaust backpressure sensor.

45. Refer to the composite vehicle (diagram on page 39 of the booklet) to answer this question: The fan will not operate from the fan control switch and the diagnostic switch will not cause the ECM to flash the DTCs. Which of the following could be the cause?

TASK B.10

   A.   A faulty fan control switch
   B.   A faulty diagnostic switch
   C.   An open at ECM pin 58
   D.   An open circuit 161

   Answer A is incorrect. A faulty fan control switch would not affect the operation of the diagnostic switch.

   Answer B is incorrect. A faulty diagnostic switch would not affect the cooling fan operation.

   Answer C is incorrect. An open at ECM pin 58 would only affect the cooling fan, not the diagnostic switch operation.

   Answer D is correct. Circuit 161 is common to both switches; if it was open, neither switch would function correctly.

PREPARATION EXAM ANSWER SHEET FORMS

ANSWER SHEET

1. _____	21. _____	41. _____
2. _____	22. _____	42. _____
3. _____	23. _____	43. _____
4. _____	24. _____	44. _____
5. _____	25. _____	45. _____
6. _____	26. _____	
7. _____	27. _____	
8. _____	28. _____	
9. _____	29. _____	
10. _____	30. _____	
11. _____	31. _____	
12. _____	32. _____	
13. _____	33. _____	
14. _____	34. _____	
15. _____	35. _____	
16. _____	36. _____	
17. _____	37. _____	
18. _____	38. _____	
19. _____	39. _____	
20. _____	40. _____	

ANSWER SHEET

1. _____	21. _____	41. _____
2. _____	22. _____	42. _____
3. _____	23. _____	43. _____
4. _____	24. _____	44. _____
5. _____	25. _____	45. _____
6. _____	26. _____	
7. _____	27. _____	
8. _____	28. _____	
9. _____	29. _____	
10. _____	30. _____	
11. _____	31. _____	
12. _____	32. _____	
13. _____	33. _____	
14. _____	34. _____	
15. _____	35. _____	
16. _____	36. _____	
17. _____	37. _____	
18. _____	38. _____	
19. _____	39. _____	
20. _____	40. _____	

ANSWER SHEET

1. _____

2. _____

3. _____

4. _____

5. _____

6. _____

7. _____

8. _____

9. _____

10. _____

11. _____

12. _____

13. _____

14. _____

15. _____

16. _____

17. _____

18. _____

19. _____

20. _____

21. _____

22. _____

23. _____

24. _____

25. _____

26. _____

27. _____

28. _____

29. _____

30. _____

31. _____

32. _____

33. _____

34. _____

35. _____

36. _____

37. _____

38. _____

39. _____

40. _____

41. _____

42. _____

43. _____

44. _____

45. _____

ANSWER SHEET

1. _____

2. _____

3. _____

4. _____

5. _____

6. _____

7. _____

8. _____

9. _____

10. _____

11. _____

12. _____

13. _____

14. _____

15. _____

16. _____

17. _____

18. _____

19. _____

20. _____

21. _____

22. _____

23. _____

24. _____

25. _____

26. _____

27. _____

28. _____

29. _____

30. _____

31. _____

32. _____

33. _____

34. _____

35. _____

36. _____

37. _____

38. _____

39. _____

40. _____

41. _____

42. _____

43. _____

44. _____

45. _____

ANSWER SHEET

1. _____

2. _____

3. _____

4. _____

5. _____

6. _____

7. _____

8. _____

9. _____

10. _____

11. _____

12. _____

13. _____

14. _____

15. _____

16. _____

17. _____

18. _____

19. _____

20. _____

21. _____

22. _____

23. _____

24. _____

25. _____

26. _____

27. _____

28. _____

29. _____

30. _____

31. _____

32. _____

33. _____

34. _____

35. _____

36. _____

37. _____

38. _____

39. _____

40. _____

41. _____

42. _____

43. _____

44. _____

45. _____

ANSWER SHEET

1. _____
2. _____
3. _____
4. _____
5. _____
6. _____
7. _____
8. _____
9. _____
10. _____
11. _____
12. _____
13. _____
14. _____
15. _____
16. _____
17. _____
18. _____
19. _____
20. _____

21. _____
22. _____
23. _____
24. _____
25. _____
26. _____
27. _____
28. _____
29. _____
30. _____
31. _____
32. _____
33. _____
34. _____
35. _____
36. _____
37. _____
38. _____
39. _____
40. _____

41. _____
42. _____
43. _____
44. _____
45. _____

Glossary

Accelerator Pedal Position (APP) Sensor The sensor that indicates the position of the accelerator pedal to the engine control module (ECM).

Actuator A device that delivers motion in response to an electrical signal.

Additive A substance added to improve or change a certain characteristic of a material or fluid.

Adsorber Catalyst An after treatment device used in diesel engines to transform NOx from exhaust emissions to nitrogen gas and H_2O (water vapor).

Aftercooler A charge air cooling device, usually water cooled.

Air-Applied A term that is usually used when referring to a fan clutch. With this style of clutch, air is applied to the clutch to engage the fan. Spring pressure holds the fan in the released (disengaged) position.

Air Compressor An engine-driven mechanism for supplying high-pressure air to the truck brake system.

Air Filter A device that captures dirt and dust and prevents them from entering the intake system.

Air-Released A term that is usually used when referring to a fan clutch. With this style of clutch, air is required to release the clutch to disengage the fan. Spring pressure holds the fan in the applied (engaged) position.

Ambient Temperature Temperature of the surrounding air. Normally, it is considered to be the temperature in the service area where testing is taking place.

Amp or Ampere Unit for measuring electrical current.

Analog Signal A voltage signal that varies within a given range (from high to low, including all points in between).

Analog-to-Digital Converter (A/D Converter) A device that converts analog voltage signals to a digital format; located in the ECM.

Antifreeze A mixture added to water to lower its freezing point.

Anti-Lock Brake System (ABS) A system that senses wheel lockup and modulates the brakes to help the operator maintain control of the vehicle.

Armature The rotating component of a (1) starter or other motor (2) generator.

ASE Automotive Service Excellence, a trademark of the National Institute for Automotive Service Excellence.

Atmospheric Pressure Weight of the air at sea level; 14.696 pounds per square inch (psi) or 101.33 kilopascals (kPa).

Backpressure The amount of restriction present in the exhaust system. Some backpressure is normal, but excessive backpressure reduces the efficiency of the engine.

Battery Terminal A tapered post or threaded studs on top of the battery case used to connect the cables.

Bimetallic Two dissimilar metals joined together that have different bending characteristics when subjected to changes of temperature.

Blade Fuse A type of circuit protection device used in automotive applications that has a plastic body and two prongs that fit into sockets.

Blower Fan A fan that pushes or blows air through a ventilation, heater, or air-conditioning system.

Bobtailing A tractor running without a trailer.

Boost Pressure Sensor A sensor mounted in the intake manifold of a diesel engine that senses the turbo boost pressure and sends that information to the engine ECM.

Camshaft Position Sensor (CMP) A sensor used to detect and communicate the position of the camshaft to the ECM.

Cavitation A condition caused by the formation and collapse of bubbles (cavities in a liquid).

C-EGR Cooled exhaust gas recirculation.

Charge Air Cooler Also known as intercooler. A device mounted in front of the radiator to cool the air leaving the turbocharger before it enters the intake manifold.

Charging Circuit The alternator (or generator) and associated circuit used to keep the battery charged and power the vehicle's electrical system when the engine is running.

Charging System A system consisting of the battery, alternator, voltage regulator, associated wiring, and the electrical loads of a vehicle. The purpose of the system is to recharge the battery whenever necessary and to provide the current required to power the electrical components.

Check Valve A valve that allows flow in one direction only.

Clutch A device for connecting and disconnecting the engine from the transmission.

Common-Rail Injector A solenoid-operated, computer-controlled diesel fuel injector that has high-pressure fuel up to 30,000 psi applied to it. The injector sprays fuel into the combustion chamber when energized by the engine's ECM.

Compression Applying pressure to a spring or fluid.

Compression Brake A device used to release the compressed air from the cylinder near the end of the compression stroke.

Condensation The process by which gas (or vapor) changes to a liquid.

Conductor Any material that permits the electrical current to flow.

Coolant Liquid that circulates in an engine cooling system.

Coolant Heater A component used to aid engine starting and reduce the wear caused by cold starting.

Coolant Hydrometer A tester designed to measure coolant-specific gravity and determine antifreeze protection.

Cooling System System for circulating coolant.

Crankcase The housing within which the crankshaft rotates.

Cranking Circuit The starter circuit, including battery, relay (solenoid), ignition switch, neutral start switch (on vehicles with automatic transmission), and cables and wires.

Crankshaft Position (CKP) Sensor Used to detect and signal the position of the crankshaft to the ECM.

Cycling (1) On-off action of the air-conditioner compressor. (2) Repeated electrical cycling that can cause the positive plate material to break away from its grids and fall into the sediment base of the battery case.

Dampen To slow or reduce oscillations or movement.

Dampened Discs Discs that have dampening springs incorporated into the disc hub. When engine torque is transmitted to the disc, the plate rotates on the hub, compressing the springs. This action absorbs the torsional vibration caused by today's low RPM, high-torque engines.

Data Links Circuits through which computers communicate with other electronic devices such as control panels, modules, sensors, or other computers; data backbone of the chassis electronic system using hardware and communications protocols consistent with CAN 2.0 and SAE J1939 standards. Also known as data bus.

Detergent Additive An additive that helps keep metal surfaces clean and prevent deposits. These additives suspend particles of carbon and oxidized oil in the oil.

Diagnostic Flow Chart A chart that provides a systematic approach to the electrical system and component troubleshooting and repair. These charts are found in service manuals and are specific to vehicle make and model.

Diagnostic Link Connector (DLC) A standardized electrical connector, usually mounted in the cab of the truck, that allows the technician to connect the scan tool to the truck's data bus.

Dial Caliper A measuring instrument capable of taking inside, outside, depth, and step measurements.

Diesel Exhaust Fluid (DEF) A urea-based fluid chemical that is used in selective catalyst reduction (SCR) systems on diesel engines to reduce NOx emissions.

Diesel Oxidation Catalyst (DOC) A muffler-like device mounted in the exhaust stream of a diesel engine to help remove pollutants.

Diesel Particulate Filter (DPF) A device mounted in the exhaust stream of a diesel engine. This device captures soot from the engine and then converts the soot to ash during periods of regeneration.

Digital Multi-Meter (DMM) Also known as a DVOM. A single tool that can be used to measure volts, ohms, and amps.

Digital Volt/Ohmmeter (DVOM) Also known as a DMM. An electrical test meter that can be used to measure voltage or resistance.

Diode An electrical device that allows current to flow in one direction, but not in the opposite direction.

Dosing Injector The injector used to spray DEF in the exhaust stream of a diesel engine equipped with selective catalyst reduction (SCR).

Driveline The propeller or drive shaft, and universal joints that link the transmission output to the axle pinion gear shaft.

Driveline Angle The alignment of the transmission output shaft, drive shaft, and rear axle pinion center line.

Driven Gear A gear that is driven by a drive gear, a shaft, or some other device.

Drive Shaft Assembly of one or two universal joints connected to a shaft or tube. It is used to transmit torque from the transmission to the differential.

Drive Train An assembly that includes all torque transmitting components from the rear of the engine to the wheels.

DTC Acronym for diagnostic trouble code. A DTC is a piece of troubleshooting data that is present when an on-board computer senses a problem.

ECM Acronym for electronic control module.

ECU Acronym for electronic control unit.

Eddy Current Circular current produced inside a metal core in the armature of a starter motor. Eddy currents produce heat and are reduced by using a laminated core.

EGR Valve Acronym for exhaust gas recirculation valve.

Electricity The movement of electrons from one location to another.

Electromotive Force (EMF) The force that moves electrons between atoms. This force is the pressure that exists between the positive and negative points; it is measured in units called volts. Charge differential.

Electronic Unit Injector (EUI) Fuel System A diesel fuel injection system that uses a camshaft lobe and rocker arm to create the force needed to inject the diesel fuel into the cylinder.

Electronically Erasable Programmable Memory (EEPROM) Computer memory that enables write-to functions.

Electrons Negatively charged particles orbiting every nucleus.

EMF Acronym for electromotive force.

Engine Brake A hydraulically operated device that converts the engine into a power-absorbing mechanism.

Environmental Protection Agency (EPA) An agency of the United States government charged with the responsibility of protecting the environment.

Exhaust Brake A slide mechanism that restricts the exhaust flow, causing exhaust backpressure to build up in the engine's cylinders. The exhaust brake transforms the engine into a power-absorbing air compressor driven by the wheels.

Failure Mode Identifier (FMI) The code used to inform the technician what type of failure has occurred in the computer-controlled system. Examples of these failures would be open, short, reading above normal, or reading below normal.

Fault Code A code stored in computer memory to be retrieved by a technician using a diagnostic tool. (See DTC)

Federal Motor Vehicle Safety Standard (FMVSS) A federal standard that specifies that all vehicles in the United States be assigned a vehicle identification number (VIN).

Feeler Gauge A tool for measuring that contains strips of metal in varying thickness, typically ranging from 0.0015 in (0.0381 mm) to 0.025 in (0.635 mm).

Fixed Value Resistor An electrical device that is designed to have only one resistance rating for controlling voltage; this rating should not change.

Foot-Pound An English unit of measurement for torque. One foot-pound is the torque obtained by a force of one pound applied to a foot-long wrench handle.

Fretting Wear that is the result of vibration between the contact points of two mating surfaces.

Fuel Cooler A device used to cool the diesel fuel; it is usually mounted in the airstream.

Fuse Link A short length of smaller gauge wire installed in a conductor, usually close to the power source.

Fusible Link A circuit protection device made of a short piece of wire with a special insulation designed to melt and open during an overload. Installed near the power source in a vehicle to protect one or more circuits, and is usually two to four wire gauge sizes smaller than the circuit it is designed to protect.

Gear A disk-like wheel with external or internal teeth that serves to transmit or change motion.

Glow Plugs Electrical heaters installed in the diesel engine combustion chamber to warm the air in order to aid cold starting and reduce white smoke. Generally used as a preheat system prior to starting the engine.

Ground The negatively charged side of a circuit. A ground can be a wire, the negative side of the battery, or the vehicle chassis.

Grounded Circuit A shorted circuit that causes current to return to the battery before it has reached its intended destination.

Harness and Harness Connectors The vehicle's electrical system wiring; it is a convenient starting point for tracking and testing circuits.

Hazardous Materials Any substance that is flammable, explosive, or known to produce adverse health effects in living beings or the environment.

Heater Control Valve A valve that controls the flow of coolant into the heater core from the engine.

High-Resistant Circuits Circuits that have an increase in circuit resistance, with a corresponding decrease in current.

Hydraulic Electronic Unit Injection (HEUI) This diesel fuel system is similar to an EUI electronic unit injection system, but instead of using a rocker arm and cam lobe, this system uses engine oil under high pressure (approximately 600–3500 psi) to inject the diesel fuel into the combustion chamber.

Hydrometer A tester designed to measure the specific gravity of a liquid.

Inboard Toward the centerline of the vehicle.

Infrared Pyrometer A tool used by the technician to determine the temperature of components.

Injector Sleeve A sleeve, located in the cylinder head, that houses the diesel fuel injector. These sleeves can be made of brass or stainless steel. Often, they can be removed and replaced while the cylinder head is still on the engine.

In-Line Fuse A fuse that is in series with the circuit in a small plastic fuse holder, not in the fuse box or panel. It is used, when necessary, as a protection device for a portion of the circuit, even though the entire circuit may be protected by a fuse in the fuse box or panel.

Insulator A material, such as rubber or glass, that offers high resistance to electron flow.

Integrated Circuit A solid-state component containing diodes, transistors, resistors, capacitors, and other electronic components mounted on a single piece of material and capable of performing numerous functions.

Intake Manifold Pressure Sensor A sensor connected to the intake manifold to measure pressure. This sensor is also known as a turbo boost pressure sensor or simply boost pressure sensor.

Jacobs Engine Brake An engine brake named for its inventor. A hydraulically operated device that converts a power-producing diesel engine into a power-absorbing retarder.

Jumper Wire A wire used to temporarily by-pass a circuit or components for electrical testing. A jumper wire consists of a length of wire with an alligator clip at each end.

Jump Start The procedure used when it becomes necessary to use a boost battery to start a vehicle with a discharged battery.

Lateral Runout The wobble or side-to-side movement of a rotating wheel.

Maintenance Manual A publication containing routine maintenance procedures and recommended service intervals for vehicle components and systems.

Manometer A tool used to measure pressure or vacuum.

NATEF Acronym for the National Automotive Technicians Education Foundation.

National Automotive Technicians Education Foundation (NATEF) An organization that provides certifying secondary and post-secondary training programs for automotive, collision repair, and heavy-duty truck technicians.

National Institute for Automotive Service Excellence (ASE) A nonprofit organization that has an established certification program for automotive, heavy-duty truck, auto body repair, engine machine shop technicians, and parts specialists.

NHTSA Acronym for the National Highway Traffic Safety Administration, a U.S. government agency.

NOP Acronym for nozzle opening pressure. The pressure at which an injector nozzle allows the spray of fuel. Also known as valve opening pressure (VOP).

OEM Acronym for original equipment manufacturer.

Off-Road A reference to unpaved, rough, or ungraded terrain on which a vehicle will operate. Any terrain not considered part of the highway system falls into this category.

Ohm A unit of electrical resistance.

Ohm's Law Basic law of electricity stating that in any electrical circuit, current, resistance, and pressure work together in a mathematical relationship.

On-Road A term that refers to a paved or smooth-graded surface on which a vehicle will operate; part of the public highway system.

Open Circuit An electrical circuit whose path has been interrupted or broken either accidentally (a broken wire) or intentionally (a switch turned off).

OSHA Acronym for the Occupational Safety and Health Administration, a U.S. government agency.

Output Driver Electronic switch that the computer uses to control the output circuit. Output drivers are located in the output ECM.

Overrunning Clutch A clutch mechanism that transmits power in one direction only.

Oxidation Inhibitor An additive used with lubricating oils to keep oil from oxidizing at high temperatures.

Parallel Circuit An electrical circuit that provides two or more paths for current flow.

Parking Brake A mechanically applied brake used to prevent a parked vehicle's movement.

Parts Requisition A form that is used to order new parts on which the technician writes the part(s) needed, along with the vehicle's VIN.

Pitting Surface irregularities resulting from corrosion.

Polarity The state, either positive or negative, of charge differential.

Pole The number of input circuits made by an electrical switch.

Pounds per Square Inch (psi) A unit of English measure for pressure.

Power A measure of work being done factored with time.

Power Flow The flow of power from the input shaft through one or more sets of gears.

Power Take-Off (PTO) An area on the engine or transmission designed to allow accessories to be mounted and powered.

Power Train A term used to describe the components from the engine to the wheels in a vehicle.

Pressure The force applied to a definite area measured in pounds per square inch (psi) English or kilopascals (kPa) metric.

Pressure Differential The difference in pressure between any two points of a system or a component.

Printed Circuit Board Electronic circuit board made of thin nonconductive material onto which conductive metal has been deposited. The metal is then etched by acid, leaving lines that form conductors for the circuits on the board. A printed circuit board can hold many complex circuits.

Programmable Read-Only Memory (PROM) An electronic memory component that contains program information specific to chassis applications; used to qualify ROM data. PROMs are used in most of the computers on cars and trucks.

Pump-Line-Nozzle Electronic (PLN-E) Fuel System A diesel engine fuel system using an in-line fuel injection pump to meter, time, and pressurize the diesel fuel. The fuel flows through an injection line to an injection nozzle. The "E" denotes that the timing and quantity of fuel delivered by the injection pump are electronically controlled.

Random-Access Memory (RAM) Memory used during computer operation to store temporary information. The microcomputer can write, read, and erase information from RAM; electronically retained.

Read-Only Memory (ROM) Memory used in microcomputers to store information permanently.

Recall Bulletin Formal notification from a manufacturer that pertains to special situations that involve service work or replacement of components in connection with a recall notice.

Reference Voltage The voltage supplied to various sensors by the computer, which acts as a baseline voltage; modified by sensors to act as input signal.

Re-Flash The procedure of erasing old software files from an ECM and installing new ones.

Refractometer A tool used to measure the specific gravity of a liquid.

Relay An electrical switch that uses a small current to control a larger one, such as a magnetic switch used in starter motor cranking circuits. It consists of a control circuit and a power circuit.

Reserve Capacity Rating The ability of a battery to sustain a minimum vehicle electrical load in the event of a charging system failure.

Resistance Opposition to current flow in an electrical circuit; measured in ohms.

Resource Conservation and Recovery Act (RCRA) A Federal law that states hazardous material must be properly stored after its use until an approved hazardous waste hauler arrives to take it to a disposal site.

Revolutions per Minute (RPM) The number of complete turns a shaft turns in one minute.

Right to Know Law A Federal law administered by the Occupational Safety and Health Administration (OSHA) that requires any company that uses or produces hazardous chemicals or substances to inform its employees, customers, and vendors of any potential hazards that may exist in the workplace as a result of using the products.

Ring Gear The gear around the edge of a flywheel.

ROM Acronym for read-only memory.

Rotary Oil Flow A condition caused by the centrifugal force applied to the fluid as the converter rotates around its axis.

Rotation A term used to describe a gear, shaft, or other device when it is turning.

RPM Acronym for revolutions per minute.

Runout Deviation or wobble of a shaft or wheel as it rotates. It is measured with a dial indicator.

Semiconductor Solid-state material used in diodes and transistors.

Sensing Voltage A reference voltage put out by the alternator that allows the regulator to sense and adjust charging system output voltage.

Sensor An electronic device used to monitor conditions for computer control requirements.

Service Bulletin Publication that provides the latest service tips, field repairs, product improvements, and related information of benefit to service personnel.

Service Manual A manual published by the manufacturer that contains service and repair information for all vehicle systems and components.

Shimming A method of adding or removing spacers for adjusting clearance or height.

Short Circuit An undesirable connection between two wires. The short occurs when the insulation is worn between two adjacent wires and the metal in each wire contacts the other or when wires are damaged or pinched.

Solenoid An electromagnet that is used to conduct electrical energy in mechanical movement.

Solvent Substance that dissolves other substances.

Spade Fuse Term used for blade fuse.

Spalling Surface fatigue that occurs when chips, scales, or flakes of metal break off.

Specialty Service Shop A shop that specializes in areas such as engine rebuilding, transmission/axle overhauling, brake, air conditioning/heating repairs, and electrical/electronic work.

Specific Gravity The ratio of a liquid's mass to that of an equal volume of distilled water.

Stall Test Test performed when there is a malfunction in the vehicle's power package (engine and transmission) to determine which component is at fault.

Starter Circuit The circuit that carries the high current flow and supplies power for engine cranking.

Starter Motor Device that converts electrical energy from the battery into mechanical energy for cranking.

Starter Safety Switch Switch that prevents vehicles with automatic transmissions from being started in gear.

Static Balance The condition of being balanced when at rest or not in motion.

Stepped Resistor A resistor designed to have two or more fixed values, available by connecting wires to one of several taps.

Storage Battery A battery to provide a source of direct current electricity for both the electrical and electronic systems.

Straightedge A piece of flat stock that has been precision ground to be flat. This tool is used in conjunction with a feeler gauge to determine if a component is warped or otherwise not flat.

Stranded Wire Wire that is made up of a number of small solid wires, generally twisted together, to form a single conductor.

Switch Device used to control and direct the flow of current in a circuit. A switch can be under the control of the driver or can be self-operating through a sensed condition of the circuit, the vehicle, or the environment.

Tachometer Instrument that indicates shaft rotating speeds.

Throttle Position Sensor (TPS or TP) A sensor that detects and communicates the position of the throttle plate to the ECM.

Throw (1) Offset of a crankshaft. (2) Number of output circuits of a switch.

Time Guide Information Used for computing compensation payable by the truck manufacturer for repairs or service work to vehicles under warranty.

Timing The phasing of events to produce action, such as ignition.

Top Dead Center (TDC) The location of the piston at its highest point in the cylinder.

Torque Twisting force.

Torque Converter A device, similar to a fluid coupling, that transfers engine torque to the transmission input shaft and can multiply engine torque.

Toxicity A statement of how poisonous a substance is.

Tractor A motor vehicle that has a fifth wheel and is used for pulling a semitrailer.

Transistor Semiconductor used to switch or amplify an electrical signal.

Tree Diagnosis Chart Chart used to provide a logical sequence for what should be inspected or tested when troubleshooting a repair problem.

Turbo-Boost Pressure Test A test that measures the pressure produced by the turbocharger. This test is usually performed with the engine under full load.

Ultra-Low Sulfur Diesel (ULSD) Fuel Diesel fuel that has a maximum of 15 parts per million sulfur.

Variable Geometry Turbocharger (VGT) A computer-controlled turbocharger that uses moveable vanes or a sliding volute to change the geometry of the turbocharger's exhaust exit area. This variability allows the ECM to control turbo boost more effectively.

VGT Actuator A computer-controlled device, operated electrically or through air pressure, that opens and closes the variable geometry turbocharger. Some VGTs may be smart devices with the ability to communicate directly over the J1939 data bus.

Vacuum Condition when pressure values are below atmospheric pressure.

Vehicle Retarder An engine or driveline brake.

Vibration Damper A round device, typically mounted to the front of the engine crankshaft, to remove the vibration caused by combustion chamber events occurring in the diesel engine. Also known as a harmonic balancer.

VIN Acronym for vehicle identification number.

Viscosity Resistance to flow or fluid sheer.

VOP Acronym for valve opening pressure; Caterpillar® term for NOP.

Volt The unit of electromotive force (EMF).

Voltage-Generating Sensors Devices that produce their own voltage signal.

Voltage Limiter Device that provides protection by limiting voltage to the instrument panel gauges to approximately five volts.

Voltage Regulator Device that controls the current produced by the alternator and thus, the voltage level in the charging circuit.

Water-in-Fuel (WIF) Sensor A device, usually located in the fuel/water separator or primary fuel filter, that alerts the vehicle's driver to a water-in-fuel condition.

Watt Measure of electrical power.

Watt's Law A principle of electricity used to calculate the power consumed in electrical circuit, expressed in watts. It states that power equals voltage multiplied by current.

Windings (1) The three bundles of wires in the stator. (2) Coil of wire in a relay or similar device.

Work (1) Forcing a current through a resistance. (2) The product of a force.

Yield Strength The highest stress a material can stand without permanent deformation or damage, expressed in pounds per square inch (psi).